CMOS DEVICES
AND TECHNOLOGY
FOR VLSI

John Y. Chen

CMOS DEVICES AND TECHNOLOGY FOR VLSI

PRENTICE HALL, *Englewood Cliffs, New Jersey 07632*

Library of Congress Cataloging-in-Publication Data

CHEN, JOHN Y.
    CMOS devices and technology for VLSI / John Y. Chen.
        p.   cm.
    Bibliography: p.
    Includes index.
    ISBN 0-13-138082-6
    1. Integrated circuits—Very large scale integration. 2. Metal
oxide semiconductors, Complimentary. I. Title
TK7874.C523 1990   88-38557
621.381'73—dc19   CIP

Editorial/production supervision
  and interior design: BARBARA MARTTINE
Cover design: DIANE SAXE
Manufacturing buyer: MARY ANN GLORIANDE

 © 1990 by Prentice-Hall, Inc.
A Division of Simon & Schuster
Englewood Cliffs, New Jersey 07632

The publisher offers discounts on this book when ordered
in bulk quantities. For more information, write:

> Special Sales/College Marketing
> College Technical and Reference Division
> Prentice Hall
> Englewood Cliffs, New Jersey 07632

Printed in the United States of America

10 9 8 7 6 5 4 3 2 1

ISBN 0-13-138082-6

PRENTICE-HALL INTERNATIONAL (UK) LIMITED, *London*
PRENTICE-HALL OF AUSTRALIA PTY. LIMITED, *Sydney*
PRENTICE-HALL CANADA INC., *Toronto*
PRENTICE-HALL HISPANOAMERICANA, S.A., *Mexico*
PRENTICE-HALL OF INDIA PRIVATE LIMITED, *New Delhi*
PRENTICE-HALL OF JAPAN, INC., *Tokyo*
SIMON & SCHUSTER ASIA PTE. LTD., *Singapore*
EDITORA PRENTICE-HALL OF BRASIL, LTDA., *Rio de Janeiro*

TO MY PARENTS

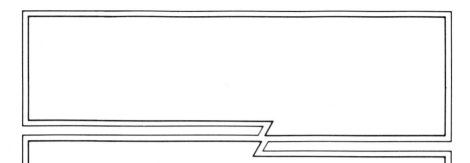

CONTENTS

4 CMOS OPERATION   92

5 CMOS PROCESS TECHNOLOGY   119

6 CMOS TRANSISTOR DESIGN   174

7 CMOS ISOLATION   233

8 LATCHUP IN CMOS   285

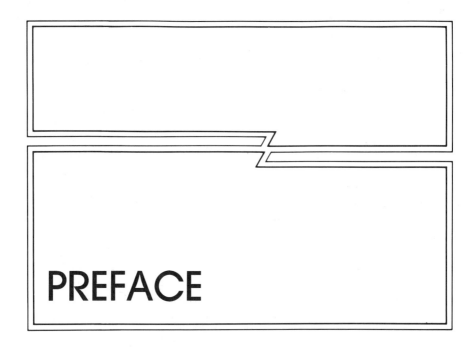

PREFACE

This book describes the device physics and IC technology of CMOS that is currently the dominant technology for VLSI. It emphasizes the means of applying device physics to technology and the impact of this technology on VLSI circuits. This book is a device technology book that bridges physics and circuits in VLSI. It will focus on device physics, yet also include process technology. Details on the interactions between device design, device modelling and process technology will be provided.

The audience of this book will include senior and graduate students in Electrical Engineering, Material Sciences, and perhaps Computer Science. The book should also be useful for professional engineers in the semiconductor and computer industries. It will teach semiconductor device engineers and process engineers/technicians how to design devices and develop CMOS processes. It will also help circuit designers to understand technology implications when implementing circuits with CMOS technology. Unlike some existing books that address the design and system aspects of CMOS circuits, this book concentrates on the device physics and process technology needed for implementing VLSI circuits and systems.

Chapter 1 Introduction to CMOS, the VLSI Technology—The material in this chapter originated from a tutorial paper titled, "CMOS—the emerging VLSI technology." I have deleted the word "emerging" from the title of this chapter for the obvious reason that CMOS is now indeed the VLSI technology. This chapter basically describes why this statement is true, and compares CMOS to NMOS, which was formerly the mainstream technology in the IC industry. Because it is only an introduction, the chapter states the facts that have made CMOS for VLSI and leaves the causes and the associated physics for the following chapters.

Chapter 2 CMOS Device Physics—This chapter reviews MOS physics to provide a background for CMOS devices and technology. The chapter's simple description follows a logical development beginning with semiconductor physics to MOS capacitors, MOS transistors, and finally discusses some effects as transistors are shrunk for VLSI. Emphases are placed on the differences between electrons and holes for the understanding of their effects on n-channel and p-channel transistors. Ideal and non-ideal MOS capacitors are introduced and threshold voltage is defined. A MOS transistor is described as two p-n junctions added to both sides of a MOS capacitor. The simple I-V relations in a MOS field-effect transistor (MOSFET) are also described. Buried-channel behaviors as well as short and narrow channel effects are considered as well. Much of the material used in this chapter is extracted from two books: Physics of Semiconductor Devices by S. M. Sze in 1981 and Physics and Technology of Semiconductor Devices by A. S. Grove in 1967. Both books are published by John Wiley & Son, Inc.

Chapter 3 MOS Modelling—This chapter describes the use of modelling tools for CMOS device design and process integration. The chapter first reviews the important models available to date. Process models such as SUPREM and SUPRA, and device models such as GEMINI, MINIMOS and PISCES are described. The basic principles, assumptions, and limitations of these models are discussed. Ways to use these models as design tools are introduced and examples are presented. Models developed to extract MOSFET parameters for circuit simulation are included. Relations among SUXES, CISM, BSIM and SPICE are discussed.

Chapter 4 CMOS Operation—CMOS operation is introduced through the description of simple CMOS circuit components. The objective of this chapter is to provide some ideas about how circuit operation is affected by the technology used. Although the emphasis is on CMOS circuits, comparisons to typical NMOS circuits are made. Basic components including inverters, buffers and transmission gates are presented first. Then, simple components for logic and memory circuits are discussed.

Chapter 5 CMOS Process Technology—This chapter discusses overall CMOS process integration. It first introduces the relationship between a circuit layout and a device cross-section at key processing steps, then describes various CMOS process technologies: p-well, n-well, twin-tub and retrograde well processes. Comparisons are made among these processes and their impacts on circuit performance are discussed. CMOS technologies for silicon on sapphire or other insulators are also described. Bipolar and CMOS integration (BiCMOS) is described extensively in the last section.

Chapter 6 CMOS Transistor Design—This chapter discusses transistor design for both n- and p- channel MOSFETs in a CMOS IC. For the sake of VLSI technology, it focuses on the design of transistors with micrometer and submicrometer channel lengths. Introduction of MOSFET scaling and non-scalable parameters are followed by the design of short-channel FETs. Hot carrier effects are examined and drain engineering is described for designing reliable nMOS transistors. Buried-channel characteristics are discussed for pMOS transistor design. Lastly, pMOS long-term reliability is studied for transistors scaled to submicrometer dimensions.

Chapter 7 CMOS Isolation—This chapter starts with the physics and techniques of field isolation in MOS technologies. What is unique in CMOS circuits is the isolation among opposite-type transistors as well as the isolation among similar-type transistors. CMOS isolation including physical and electrical isolation, failure mechanisms, and dependence on process technology are discussed and associated design rules are derived. New isolation techniques for future CMOS are also presented.

Chapter 8 Latchup in CMOS—A major problem in CMOS, known as latchup, is a strong function of circuit density, nMOS to pMOS separation in particular. Both device technologists and circuit designers should have knowledge of this subject for developing CMOS. This chapter covers the physics and modelling of latchup, its characterization in steady state and transient, and various methods of avoiding latchup. The author would like to acknowledge that much of the material presented here is borrowed from the journal articles and internal reports published by Mr. Alan Lewis *et al.* Discussions with Mr. Alan Lewis and Dr. Ron Troutman are also acknowledged.

Chapter 9 CMOS Design Rules—In this last chapter, design rules associated with current and future CMOS technologies are discussed. Because design rules are derived from device characteristics and process capabilities, it is adequate to summarize the device parameters discussed in the previous chapters, then to discuss the effect of particular process(es) on individual design rules. Lambda-based design rules and the limitation of the rules are discussed. Modification of lambda rules to accommodate technology needs are also presented. Examples are given for one-to-two micrometer CMOS design rules. Extending to submicrometer rules and related issues are described.

This book originated from my lecture notes for a graduate course, "CMOS for VLSI," taught at the Electrical Engineering Department at Santa Clara University in 1985 and 1986. I would like to thank Professor Cary Yang for accommodating my working schedule by allowing me to teach this course at the "early bird" program. Even before this opportunity, I was motivated to start this project by the overwhelming response of the CMOS short courses organized by Prof. C. R. Viswanathan and myself through UCLA in 1984. My knowledge and understanding of this topic have certainly improved by giving lectures and participating in other lectures offered by distinguished speakers such as Dr. Tony Alvarez, Dr. Pallab Chatterjee, Dr. Wolfgang Fichtner, Dr. Ken Martin, Dr. Lou Parrillo, Dr. Ranjeet Pancholy, Dr. Charles Sodini, Dr. Ron Troutman and Dr. Ken Yu.   I was also encouraged by receiving the Outstanding Paper Award for my review paper on "CMOS— the emerging VLSI technology," given at the International Conference on Computer Design: VLSI in Computer.   I thank Drs. Lou Parrillo and Jim Clemens for inviting me to give the paper.

I am deeply indebted to Dr. S. T. Chang at HP Lab for his critical reading of Chapter 2, Prof. Ping Ko at U.C. Berkeley for his comments on Chapter 3, Prof. Charles Sodini at MIT for his suggestions on Chapter 4, Prof. Bruce Wooley at Stanford Univ. for his reading and comments on Chapter 5, especially the BiCMOS section, Prof. Chenming Hu at U.C. Berkeley for his review on Chapter 6, Dr. Chi Chang at AMD for discussions in Chapter 7, Dr. Ron Troutman at IBM and Mr. Alan Lewis at Xerox for their assistance in Chapter 8, and Dr. George Lewicki at ISI/MOSIS for looking over Chapter 9.   Comments provided by the reviewers through my publisher were also extremely useful. Finally, I would like to thank Prof. David Hodges at U.C. Berkeley who has always encouraged me in my writing and teaching.

In addition, I thank my secretary, Judy Quinn, and my wife, Heather who helped a great deal in polishing the final manuscript.   At my publisher, Prentice-Hall, I want to acknowledge Mr. Bernard Goodwin and Mr. Dan Joraanstad who encouraged me to undertake this project, and Ms. Barbara Marttine who handled the production of this book.   Lastly, I wish to thank my parents Chiang-Ho and Pei-Hwan Chen, and my wife, Heather for their continuous encouragement and support.

JOHN Y. CHEN

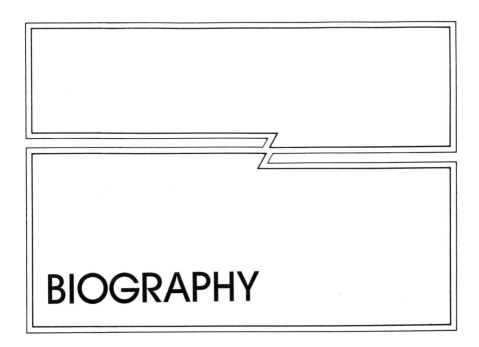

BIOGRAPHY

Dr. John Y. Chen has 13 years of experience in IC technology and eight years in CMOS. He was a Howard Hughes Doctoral Fellow and received a Ph.D. from UCLA in Electrical Engineering. At Hughes Research Laboratories (HRL), he has worked on various VLSI technologies for radiation-hard CMOS. He received the HRL Outstanding Paper award and several invention Awards. At Xerox Palo Alto Research Center, he led a team in developing and transferring 1.2 micron CMOS technology for custom ICs. He received Xerox Team Excellence award and Xerox Achievement award in 1985. Dr. Chen is currently at the Microelectronics Laboratory of Boeing High Technology Center and leads projects such as BiCMOS and GaAs CMES. He also gives CMOS lectures at the EE department of University of Washington, Seattle.

Dr. Chen served as a technical program committee member at the International Electron Device Meeting (IEDM) for three years and chaired the Solid-State Device Sub-Committee in 1986. He has also been the committee member for the Symposium on VLSI Technology since 1986. He served as the Technical Program Chairman for the 1989 International VLSI Symposium on Technology, Systems and Applications.

He has organized and lectured several short courses on CMOS at UCLA and the University of Maryland. He taught a course on CMOS Device and Technology at the EE department of Santa Clara University. He has given graduate-level seminar lectures at the University of California-Berkeley and Stanford University. He has authored and presented 80 technical papers of which more than half related to CMOS. He has also written a chapter for the book, *Advanced MOS Device Physics*, published by Academic Press in 1988. He has several patents on CMOS devices and technology. His invited paper, "CMOS—The Emerging VLSI Technology," won Outstanding Paper Award at the 1985 International Conference on Computer Design: VLSI in Computer.

CMOS DEVICES
AND TECHNOLOGY
FOR VLSI

1

INTRODUCTION TO CMOS, THE VLSI TECHNOLOGY

Although CMOS logic was introduced in the early 1960s,[1] its applications were limited to special circuits, such as watches and calculators, which have very low standby power or radiation-hard circuits that require very high noise margin. In the 1960s and 1970s, CMOS in general compared unfavorably with NMOS because of its lower speed and poorer packing density. It also required more complicated fabrication processes and special layout attention for latchup prevention.

The 1980s is known as the decade of VLSI. VLSI is the integration of *many small* transistors on a single chip. With transistor counts of about a million or more, power dissipation becomes a fundamental limitation. Consequently, CMOS technology is gaining a great deal of attention in VLSI because of its inherent low power characteristics. Shrinking transistor size for VLSI implementation is another driving force for CMOS because, as transistor dimensions are reduced, the current delivered by a *p*-channel transistor approaches the current provided by an *n*-channel device of the same size. As a result, small CMOS devices are not much slower than their NMOS counterparts. VLSI speed depends more on the actual circuit design rather than the moderate difference in device driving ability. Indeed, some CMOS circuits have shown comparable or better speed than corresponding NMOS circuits due to relaxed power limitation. For example, the 64K CMOS DRAM made by Intel Corporation has 70 ns access time and a standby power of 25 mW,[2] whereas Intel's DRAM made using NMOS technology has approximately the same column access time, but much higher power dissipation.[3]

In the early days, the CMOS process had more masking steps and was more complicated than NMOS because two different devices must be made on the same chip. Again, as MOS technology was scaled down to one micrometer, NMOS processes evolved to more complexity because of the need for punchthrough protection, multiple threshold voltages, buried contacts and/or polysilicon resistors. On the other hand, CMOS complexity has remained basically unchanged or improved slightly due to various process innovations. The net result is that the difference of process complexity between CMOS and NMOS is marginal. Moreover, as the design cost increases rapidly for VLSI, process complexity makes much less impact on total cost. The turnaround time, however, may be a much more important aspect. This feature is especially critical for ASICs (Application Specific Integrated Circuits).[4]

A CMOS logic gate uses more silicon layout area than the equivalent NMOS gate because a larger p-channel transistor is required for the load to compensate for its lower mobility. Additionally, more isolation space is needed between the two different types of transistors to avoid field leakage and latchup. To date, CMOS has lower packing density than does NMOS. However, as the integration level increases, a larger portion of chip area is dedicated to interconnects. This ratio is particularly apparent in random logic design, in which chip area is not strongly affected by the gate size. Even in SRAMs, where memory cells represent roughly 70% of total chip area, new CMOS technologies such as retrograde well, epitaxy and trench isolation have demonstrated very dense circuits.[5]

From the standpoint of circuit design, CMOS technology is easy for both digital and analog applications. With the p-channel transistor as a dynamic load device, the static operation of a CMOS inverter does not depend on transistor width ratio, thereby providing high noise immunity. This type of ratioless logic is extremely important when designing gate arrays and standard cells that must be flexible because transistors are laid out with no precise definition of the actual circuit. Although pure-static CMOS requires multiple input gates, dynamic circuit designs such as Domino CMOS[6] eliminate this problem. Dynamic circuit designs can also be used to maximize speed and minimize transistor counts with minimum effect on power dissipation. Furthermore, CMOS shows an advantage in clock generation because it can be implemented by a simple inverter chain. The number of clocks can be reduced. For example, in Intel's DRAM address buffer circuit, clocks and transistor counts have been reduced by a factor of two by replacing NMOS with CMOS.[2] In the analog world, CMOS outperforms NMOS because its high noise immunity and low power are essential to designing high-gain operational amplifiers and other analog building blocks. Bilateral switches also work much better with CMOS transmission gates because n- and p-channel FET's can be paired to eliminate body effect as observed in NMOS.

CMOS can enhance circuit reliability in various aspects. It has been shown that the soft error rate (SER) in a DRAM can be reduced dramatically when memory arrays are made inside a well, with opposite polarity to the substrate.[3] The reason is that a reversely biased p-n junction (well-substrate in CMOS) provides a potential barrier against carriers generated in the substrate. This configuration is inherent in CMOS technology.

It is known that the high electrical field in short n-channel transistors generates hot electrons which then cause high substrate current and long-term threshold shift.[7] CMOS can ease this problem by several means. First, because hole impact ionization is very low, p-channel load transistors in a CMOS logic suffer much less from hot-carrier induced degradation than n-channel depletion devices in NMOS circuits do. Second, CMOS circuits do not draw static current thereby reducing hot-carrier induced device degradation. Third, unlike NMOS circuits, CMOS does not often use bootstrapping or a substrate-bias technique which increases the electrical field in a device and may further aggravate the hot-electron problem. Finally, when CMOS is used for DRAMs, p-channel devices can be placed in memory arrays to avoid impact ionization current, thus maintaining charge storage.[2] Electromigration is another major concern in VLSI reliability; it is expected that CMOS suffers less from electromigration because no static current flows in metal lines.

In summary, CMOS, when compared with NMOS, has much lower power, comparable speed, slightly poorer density but better reliability. The major advantage to using CMOS is power dissipation. In a VLSI circuit, too many devices exist and they run very fast. As a result, more power is dissipated from each device and even more power is dissipated from a chip. Current packaging technology limits power dissipation at a few watts per chip to avoid high temperature induced reliability problems. Sophisticated packages are being developed, but at great expense. CMOS, with its inherent low power characteristics may be the only VLSI technology which can live up to thermal packaging limitations.

REFERENCES

1. G. E. Moore, C. T. Sah and F. M. Wanlass, "Metal-oxide-semiconductor field-effect devices for micropower logic circuitry," in *Micropower Electronics*, Ed. by E. Keonjian, Pergamon Press, p. 41, 1964.
2. R. J. C. Chwang, M. Choi, D. Creek, S. Stern, P. H. Pelley, III, J. D. Schutz, P. A. Warkentin, M. T. Bohr, and K. Yu, "A 70 ns high density 64K CMOS dynamic RAM," *IEEE J. of Solid-State Circuits*, vol. SC-18, p. 457, 1983.

3. P. Madland, J. Schutz, R. Green, R. King, "CMOS vs. NMOS comparisons in dynamic RAM design," in the *Proc. of ICCD*, p. 379, 1983.

4. R. W. Broderson, "The technological requirements for application specific integrated circuits," in *IEDM digest*, p. 13, 1985.

5. K. Hashimoto, Y. Nagakubo, S. Yokogawa, M. Kakumu, M. Kinugawa, K. Sawada, T. Sakurai, M. Isobe, J. Matsunaga and T. Izuka, "Deep trench well isolation for 256 Kb 6T CMOS static RAM," in the *Tech. Dig. Symp. on VLSI Technology*, p. 94, 1985.

6. B. T. Murphy and R. Edwards, "A CMOS 32b single-chip microprocessor," in *ISSCC Tech. Dig.*, p. 230, 1981.

7. T. H. Ning, P. W. Cook, R. H. Dennard, C. M. Osburn, S. E. Schuster, and H. N. Yu, "1 μm MOSFET VLSI technology: Part IV:Hot-electron design constraints." *IEEE Trans. Electron Devices*, vol. ED-26, pp. 346–353, 1979.

2

CMOS DEVICE
PHYSICS

This chapter reviews MOS physics. It should form the basis necessary for CMOS device design and technology. This chapter uses a simple description and follows the logical development from semiconductor physics to MOS capacitors, MOS transistors, and finally some important effects as transistors are shrunk for VLSI. Emphasis is placed on the differences between electrons and holes for the understanding of n-channel and p-channel transistors, which will be discussed in later chapters.

This chapter starts with a discussion of mobility and impact ionization for electrons and holes because these two basic phenomena determine CMOS transistor performance such as current driving ability and substrate current. Ideal and non-ideal MOS capacitors are introduced and the definition of threshold voltage is presented. Next, a MOS transistor is described by considering two p-n junctions added to both sides of a MOS capacitor. The simple I-V relation in a MOS field-effect transistor (MOSFET) is described and a discussion of its operations in linear, saturation and subthreshold regions follows. In addition, buried-channel behaviors are discussed because one of the two devices in a typical CMOS IC behaves this way. Short and narrow channel effects associated with small devices are also considered.

2.1 ELECTRONS AND HOLES

In semiconductors, current conduction is caused by the movement of charged carriers. In an absolutely pure (intrinsic) semiconductor, say silicon (Si), electrons are negatively charged carriers which are created due to Si bond breakage caused by thermal vibration of the Si atoms. When an Si bond is broken, a free electron results, thus leaving a vacancy in the original bond (Fig. 2.1a). The vacancy is positively charged, and is referred to as a "hole". An electron can hop from one neighboring bond into the hole and cause another hole to occur. This current can be viewed as the movement of positively charged holes in the opposite direction. The free carrier (electron or hole) density in an intrinsic Si is defined as n_i which is 1.45×10^{10} cm^{-3} at room temperature.[1]

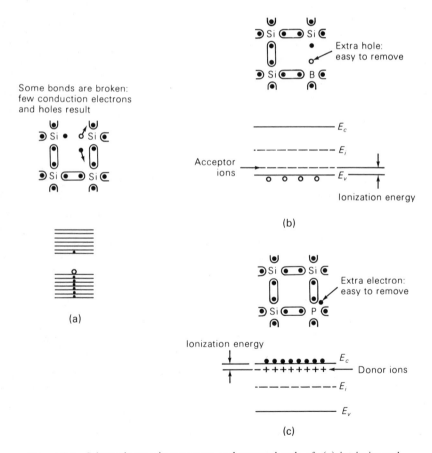

Figure 2.1 Schematic atomic structures and energy bands of: (a) intrinsic; and (b) extrinsic Si (Ref. 1).

In a doped (extrinsic) semiconductor, electrons or holes can be generated by introducing impurity atoms such as phosphorus or boron into Si. The doped Si is then called n- or p-type respectively. Compared to Si, phosphorus or arsenic has one extra valence electron; this type of atom is called a donor impurity. Boron has one less valence electron, i.e., one hole in the Si lattice; this type of atom is called an acceptor impurity. The extra electron or hole can easily escape from the Si bond and become a free carrier responsible for current conduction. It leaves the impurity atom ionized (Fig. 2.1b) in this situation.

Sufficient thermal energy exists at room temperature to ionize all the donor or acceptor impurity atoms in Si. As a result, the electron density (n) in an n-type Si is the same as the donor impurity concentration (N_D), and the hole density (p) in a p-type Si is equal to the acceptor impurity concentration (N_A). Another way to create free electrons or holes is by applying a voltage to Si to bend the Si band to a degree that electrons or holes can be locally generated. It can be shown[1] in any case, that $np = n_i^2$ for semiconductors at equilibrium.

For CMOS devices, the two most important properties of electrons and holes are carrier mobility and impact ionization. In a MOSFET, mobility affects the drain current and impact ionization determines the substrate current. Moreover, these two properties are quite different for electrons and holes.

2.1.1 Mobility

When an electrical field is applied to a semiconductor, carriers (electrons or holes) are accelerated and also scattered by lattice vibration and impurity collision. Consequently, a constant carrier velocity defined as drift velocity results. The drift velocities for electrons and holes in Si are dependent on the electrical field as shown in Fig. 2.2. At a low field, the velocities are proportional to the electrical field, hence constant mobilities (μ) can be defined as the average velocities per unit electrical field (v/E). Notice that in Fig. 2.2a the electron mobility is about 3–4 times higher than the hole mobility at low field, i.e., $E < 10^4$ V/cm. This phenomenon has resulted in more drain current in an n-channel MOSFET than that in a corresponding p-channel device. However, at a large field (about 10^5 V/cm or above), electron and hole velocities approach a common asymptote as shown in Fig. 2.2b.[2] Thus, for short-channel MOSFETs in which carriers experience high electric field and approach saturated velocities, the difference in current between an n- and a p-channel device decreases greatly.

Although carrier mobility in bulk Si is a function of impurity doping concentration,[1] mobility of charges in a MOSFET channel is not dependent on impurity concentration unless the concentration is very large, $N_A > 1.0 \times 10^{17}$ cm^{-3}.[3] Experiments have verified this observation and shown only

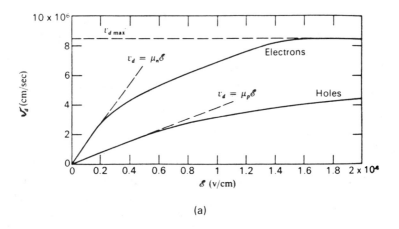

(a)

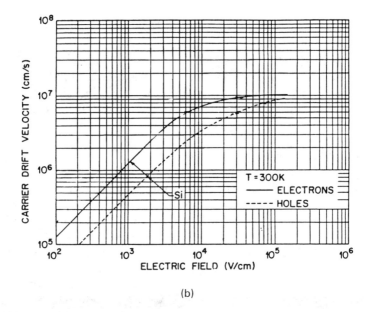

(b)

Figure 2.2 Carrier drift velocities for Si in: (a) low electric fields (Ref. 1); and (b) a wide range of electric fields (Ref. 2).

<5% difference in mobility values among a large range of doping levels. For commonly used channel doping levels, channel mobility μ_{ch} is mainly a function of the effective electric field normal to the Si surface.[3,4] It can be described empirically as[5]

$$\mu_{ch} = \mu_o(E_{crit}/E_{eff})^{UEXP} \qquad (2.1)$$

where μ_o is the mobility at the electric field of E_{crit} or lower.   UEXP and

E_{crit} are constants obtained empirically. Channel mobility decreases as the effective electric field (E_{eff}) exceeds the critical field (E_{crit}). E_{eff} is described in the discussion of MOSFET operation (see Section 2.3.2).

2.1.2 Impact Ionization

When the electric field in a semiconductor is increased to 10^5 V/cm or above, electrons or holes gain enough energy to excite and generate electron-hole pairs by impact ionization. The generation rate is given by

$$G = \alpha_n n v_n + \alpha_p p v_p \tag{2.2}$$

where α_n is the electron ionization rate defined as the number of electron-hole pairs generated by an electron per unit distance traveled; α_p is the hole

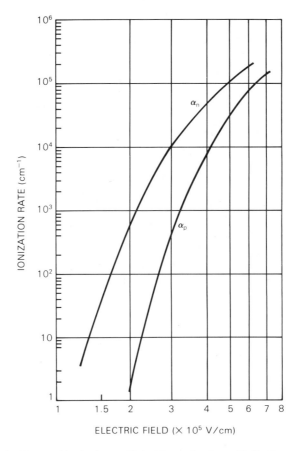

Figure 2.3 Measured ionization coefficient for avalanche multiplication vs. electric field for Si (Ref. 2).

ionization rate defined similarly; n and p are the electron and hole concentrations; and v_n and v_p are the thermal velocities for electrons and holes, respectively. The ionization rates in Si are strongly dependent on the electrical field and can be defined as:

$$\alpha_n = 3.8 \times 10^6 \exp\left(-1.75 \times 10^6/E\right) \quad (2.3a)$$

$$\alpha_p = 2.25 \times 10^7 \exp\left(-3.26 \times 10^6/E\right) \quad (2.3b)$$

The ionization rates are shown in Fig. 2.3 for electrons and holes in Si at room temperature.[2,6] They are known to decrease as the temperature increases.[7] The ionization rate for electrons is higher than that for holes by one-to-two orders of magnitude. This difference causes the substrate current in an n-channel MOSFET to be several orders of magnitudes higher than that in a p-channel device.

2.2 MOS CAPACITORS

The simplest device in MOS technology is the MOS capacitor, consisting of Metal (or other conductors such as doped polysilicon), Oxide and a Semiconductor as shown in Fig. 2.4. It is different from the capacitor made by two conducting parallel plates because the MOS capacitance is strongly dependent on the voltage applied on the gate. Electron energy band diagrams

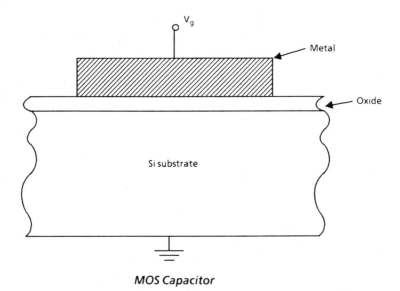

MOS Capacitor

Figure 2.4 A MOS (Metal Oxide Silicon) capacitor.

of MOS capacitors with a p-type and an n-type semiconductor are shown in Fig. 2.5 for three gate bias conditions. E_C and E_V are the conduction and valence band edges. E_i, the intrinsic level, is at the middle of the energy band gap, i.e., $E_i = (E_C + E_V)/2$. It corresponds to the Fermi level of an intrinsic-type (i.e., undoped) semiconductor. For a doped semiconductor, the Fermi level is closer to the conduction band for an n-type semiconductor, but closer to the valence band for a p-type semiconductor. In equilibrium, the Fermi level (E_F) is constant in all cases because no current conducts.

Consider the p-type semiconductor first. If a negative voltage is applied on the gate (Fig. 2.5a), the band bends up causing majority carriers (holes in this case) to accumulate near the semiconductor surface. In this accumulation mode, the MOS capacitor behaves just like a parallel-plate capacitor

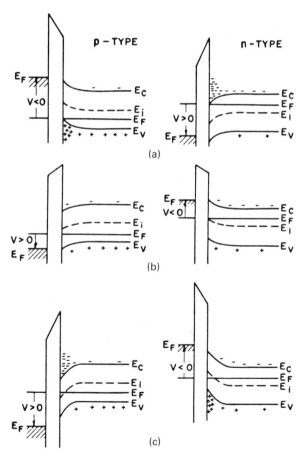

Figure 2.5 Energy band diagrams for ideal MOS capacitors for: (a) accumulation; (b) depletion; and (c) inversion (Ref. 2).

with a constant capacitance per unit area (C_{ox})

$$C_{ox} = \varepsilon_{ox}/t_{ox} \qquad (2.4)$$

where ε_{ox} and t_{ox} are the dielectric constant and thickness of the oxide. If a small positive gate voltage is applied (Fig. 2.5b), the band bends downward and the holes are pushed away from the surface, leaving a depletion region consisting of negatively charged acceptor ions. In this depletion mode, the MOS capacitor has two capacitances in series, the oxide capacitance (C_{ox}) and the depletion capacitance (C_d). While C_{ox} is constant

$$C_d = \varepsilon_{si}/d \qquad (2.5)$$

where d, the depletion width, increases as the gate voltage (V_g) is increased. The corresponding bands are bent downward more and eventually the intrinsic level E_i at the surface crosses over the Fermi level E_F (Fig. 2.5c). At this point, the semiconductor surface is inverted from p-type to n-type. Further increase of V_g will pull the bands down continuously until the E_i at the Si surface is one Φ_B below E_F, where Φ_B, bulk potential, is defined as $(E_i - E_F)$ in bulk Si. This condition is referred to as strong inversion and the corresponding band diagram is shown in Fig. 2.6.

Additional gate voltage will attract more electrons to the surface but will not widen the depletion region significantly. This instance holds true because once the onset of strong inversion has occurred, a marginal increase in a band bending due to the expansion of the depletion width results in a very large increase in the number of electrons within the inversion layer.[2] Because the depletion region remains more or less unchanged after strong inversion is reached, a maximum depletion width (d_{max}) can be defined. As shown in Fig. 2.6, the potential Φ is measured as the amount of band bending with respect to the intrinsic level (E_i) in bulk Si. At the semiconductor surface, $\Phi = \Phi_s$, and Φ_s is called the surface potential. Notice $\Phi_s = 2\Phi_B$ at strong inversion. Electron and hole concentrations can be expressed as

$$n = n_i \exp(E_F - E_i/kT) \qquad (2.6a)$$

$$p = n_i \exp(E_i - E_F/kT) \qquad (2.6b)$$

Note that $np = n_i^2$ was previously described. At strong inversion, the electron concentration at the surface is equal to the hole (majority carrier) concentration in the bulk. In the strong inversion mode, the positive charges on the gate (Q_m) must be balanced out by the electrons at the inversion layer (Q_n) and the negative charges in the depletion region (Q_B)

$$Q_m = -(Q_n + Q_B) \qquad (2.7)$$

where $Q_B = -qN_A d_{max}$. The total capacitance is the oxide capacitance in series with the depletion capacitance at the maximum depletion width in equilibrium, i.e. $C_{dmax} = \varepsilon_{si}/d_{max}$. The above discussion is for a p-type

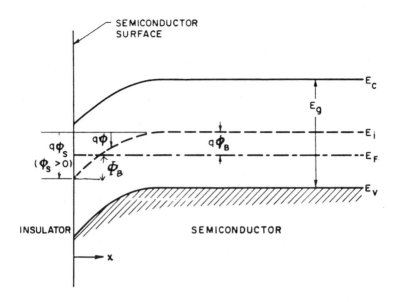

Figure 2.6 Energy band diagram at the surface for a *p*-type semiconductor (Ref. 2).

semiconductor. Similar results can be obtained for an *n*-type semiconductor, however the polarity of the voltages and charges must be reversed.

Fig. 2.7 shows the C-V characteristics of a MOS capacitor with a *p*-type substrate. The capacitance at negative voltages is constant and is equal to C_{ox} because the device is in the accumulation mode. As the gate voltage (V_g) approaches zero and becomes positive, three curves represent low-frequency (curve a), high-frequency (curve b) and pulse measurements (curve c).

Consider the high frequency case first, i.e., curve (b). When a positive voltage is applied, holes are pushed away leaving a negatively charged depletion region. At high enough frequencies, charge increment only occurs at the depletion edge. Thus, the depletion capacitor is in series with the oxide capacitor forming the MOS capacitance per unit area as

$$C = (1/C_{ox} + 1/C_d)^{-1} \qquad (2.8)$$

The higher the V_g, the larger the depletion width. And, from Eq. 2.5, C from C_d reduces as V_g is increased. Thus, the total capacitance, C, finally levels off because the maximum depletion width is reached. When the measurement frequency is low (curve a), the generation-recombination rate can keep pace with the small AC signal variation. Hence, the minority carriers (electrons, in this example) can modulate the inversion charges at the semiconductor surface. Consequently, the capacitance measured at inversion mode will be the oxide capacitance.

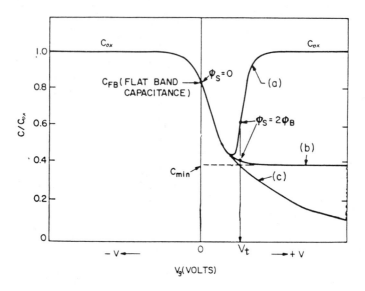

Figure 2.7 MOS capacitance-voltage curves. (a) Low frequency; (b) high frequency; and (c) deep depletion (Ref. 2).

For both low- and high-frequency cases, the DC-bias of the MOS system is swept at sufficiently low rates so that at least a quasi-equilibrium condition is maintained. However, if a voltage pulse is suddenly applied to the gate (curve c), holes are pushed away immediately and all incremental charge appears at the edge of the depletion region. Sufficient time does not exist for the generation-recombination process to create minority carriers (electrons) at the surface and supply holes to neutralize ionized acceptors at the depletion edge. This condition is non-equilibrium and is often referred to as deep depletion, which is the operational condition for charge-coupled devices (CCDs). In this case, maximum depletion width cannot be defined.

The gate voltage, however can be given as

$$V_g = Q_m/C_{ox} + \Phi_s \qquad (2.9)$$

where the first term corresponds to the voltage across the oxide and the second term represents the voltage drop in the semiconductor. At the onset of strong inversion

$$Q_m = -Q_B = qN_A d_{max} \qquad (2.10a)$$

$$\Phi_s = 2\Phi_B = qN_A(d_{max})^2/(2\varepsilon_s) \qquad (2.10b)$$

And the gate voltage (V_g) at this condition is defined as the threshold voltage, V_t. From the above two equations, V_t can be expressed as

$$V_t = 2\sqrt{\varepsilon_s q N_A \Phi_B}/C_{ox} + 2\Phi_B \qquad (2.11)$$

This equation describes the threshold voltage for an ideal MOS capacitor, defined as a MOS capacitor that gives no band bending at zero gate voltage. However, it is often unrealistic in a non-ideal MOS capacitor. The difference between the work functions of a gate and a semiconductor can lead band bending at $V_g = 0$ V. The work function is defined as the minimum energy needed for an electron to escape from its Fermi level into vacuum.

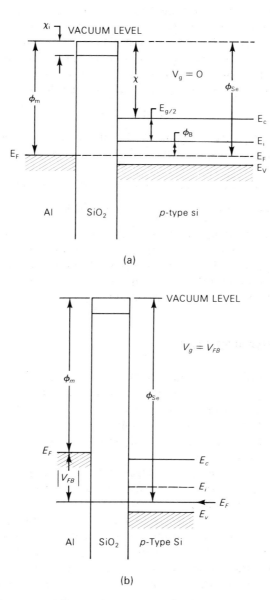

Figure 2.8   Energy band diagrams for: (a) an ideal MOS, i.e., no work function difference; and (b) a non-ideal MOS, i.e., with work function difference (Ref. 1).

To avoid band bending so that a flat-band condition can be achieved, an additional gate voltage must be applied to counterbalance the difference in work functions. Fig. 2.8 shows the energy band diagrams for an ideal MOS capacitor and an MOS capacitor with a work function difference balanced by a non-zero flat-band voltage. Φ_m is the work function of the metal gate and the work function for the semiconductor is Φ_{se}, which is equal to $\chi + E_g/2 + \Phi_B$ for a p-type semiconductor. χ is the electron affinity, defined as the energy difference between the conduction bend and the vacuum level. The charges in the oxide can induce an image charge in the semiconductor. Again, to obtain a flat-band condition, an opposite voltage must be applied to the gate. The total gate voltage needed to offset the work function difference and the oxide charge is called flat-band voltage and is given as

$$V_{FB} = \Phi_{ms} - Q_{fc}/C_{ox} \tag{2.12}$$

where $\Phi_{ms} = \Phi_m - \Phi_{se}$, Φ_m and Φ_{se} are the work functions for the gate and the semiconductor, respectively, and Q_{fc} is the fixed charge density of the oxide.

The threshold voltage for a non-ideal MOS capacitor is then written as

$$V_t = \Phi_{ms} - Q_{fc}/C_{ox} + 2\sqrt{\varepsilon_s q N_A \Phi_B}/C_{ox} + 2\Phi_B \tag{2.13}$$

In practical device design and fabrication, a thin layer of charges is commonly introduced at the silicon surface to adjust threshold voltage in a way similar to the effect of Q_{fc} in Eq. 2.13. This introduction is normally accomplished with ion implantation of boron or arsenic at a very low energy level. The ion implantation process is described in Chapter 3.

2.3 MOS TRANSISTORS

The MOS capacitor discussed previously can be made into a transistor if two p/n junctions are added at both sides of the capacitor. Fig. 2.9 shows the schematic of an n-channel MOS field-effect transistor consisting of a MOS capacitor and two n^+ regions in a p-type Si substrate. It has four terminals: gate for the capacitor, source and drain for the two n^+, and substrate (or body) for the p-substrate.

The gate can be made by a layer of metal (aluminum is most common) or heavily-doped polysilicon. Today, polysilicon gates are commonly used because they allow the alignment of the two n^+ regions (source and drain) to the edges of the gate. This self-aligned feature cannot be achieved with an aluminum gate because of process incompatibility during fabrication.

The two n^+ regions are isolated by the p-type substrate and no current can flow unless the surface of p-substrate is inverted by an adequate voltage being applied on the gate. At this point, an n-type inversion layer is formed, called an n-channel, connecting source and drain so that a positive voltage

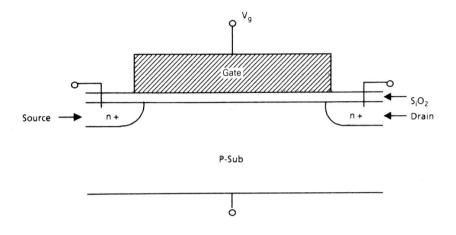

MOS FET

Figure 2.9   A MOS Field Effect Transistor (FET).

on the drain can attract electrons to flow from source to drain. The current starts to increase linearly with the drain voltage, but eventually saturates at a large drain voltage.

2.3.1 Band Structure in Non-equilibrium

Consider the band structure near the drain junction under the gate. At zero drain voltage ($V_d = 0$), only a built-in potential (V_{bi}) exists between the *p-n* junction to counter-balance the diffusion force; the *p-n* junction is in equilibrium and the Fermi level is constant through the channel direction, i.e. *y*-axis. When a gate voltage (V_g) is applied with $V_g > V_t$, the band bends $2\Phi_B$ (i.e., $\Phi_s = 2\Phi_B$) making the conduction band (E_c) closer to E_F, resulting in surface inversion as described earlier. This instance is shown in Fig. 2.10a.

Now, if a positive V_d is applied at the n^+ of the *p-n* junction, the reverse bias will create a larger potential barrier ($V_{bi} + V_d$) across the *p-n* junction, thus lowering the quasi-Fermi level for electrons (E_{Fn}) by V_d and placing the system in non-equilibrium. Therefore the band bending is not enough to bring E_c near E_{Fn} to cause inversion (Fig. 2.10b). Higher V_g must be applied to bend the band by ($2\Phi_B + V_d$) and lower the E_c under the gate closer to E_{Fn}. Fig. 2.10(c) shows this condition. The gate voltage for inversion to occur in a MOSFET channel is now a function of V_d.

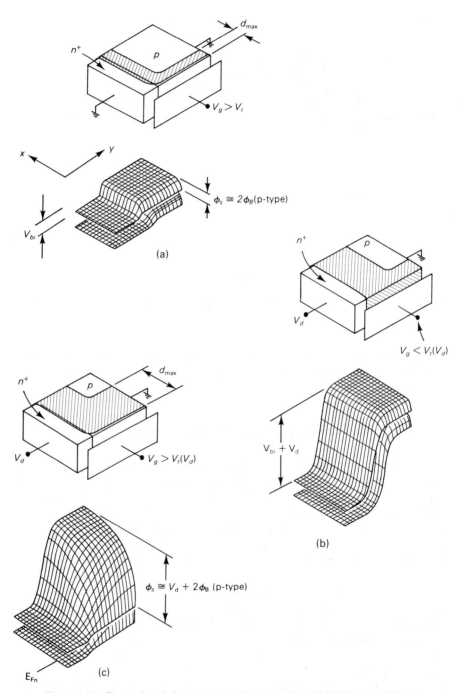

Figure 2.10 Energy band diagrams near the drain of a MOSFET with different gate and drain bias conditions; (a) $V_g > V_t$, $V_d = 0$, (b) $V_g < V_t (V_d)$, $V_d > 0$; and (c) $V_g > V_t (V_d)$, $V_d > 0$ (Ref. 1).

2.3.2 MOSFET Operation

When a voltage greater than V_t is applied to a gate, the semiconductor surface is inverted to n-type and a MOSFET channel is formed, as shown in Fig. 2.11. For small drain voltages, the entire channel is inverted and channel conductance is proportional to the inversion charges. This area is the linear region in which the drain current I_d is proportional to V_d and the FET acts as a resistor. As V_d increases and reaches a point such that $(V_d + 2\Phi_B)$ is just larger than the band bending produced by the gate voltage, the inversion disappears at point X near the drain (Fig. 2.11b). The channel is then pinched off at this point and the corresponding drain voltage is defined as V_{dsat} because for $V_d > V_{dsat}$, I_d remains essentially constant. As shown in Fig. 2.11(c), the additional voltage above V_{dsat} is consumed for widening the depletion region and the potential at Point X remains constant. Point X moves toward the source as V_d exceeds V_{dsat}, but the movement is very slight; hence, I_d increases very little. The drain current basically remains at a constant level, I_{dsat}, for $V_d > V_{dsat}$. However, for a device with very short channel length, a slight movement of the pinchoff point can be a significant portion of the entire channel length, thereby causing significant increase of I_{dsat}. Other short channel effects will be discussed in Sec. 2.5.

Linear and saturation regions. This section derives the basic MOSFET characteristics. If a MOSFET is biased with sufficient gate voltage to cause surface inversion, the free charge in the inversion layer at a distance y from the source is

$$Q_n(y) = Q_s(y) - Q_B(y) \tag{2.14}$$

where $Q_s(y)$ is the charge induced in the semiconductor per unit area at Point y. Because

$$Q_s(y) = -C_{ox}[V_g' - \Phi_s(y)] = -C_{ox}\{V_g' - [2\Phi_B + V(y)]\} \tag{2.15}$$

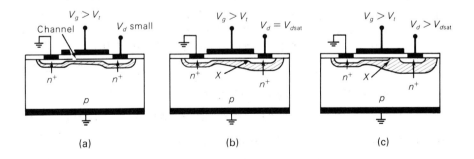

(a)   (b)   (c)

Figure 2.11 Illustration of the operation of a MOSFET for $V_g > V_t$: (a) $V_d \ll V_{dsat}$; (b) $V_d = V_{dsat}$; and (c) $V_d > V_{dsat}$ (Ref. 1).

where $V_g' = V_g - V_{FB}$, and

$$Q_B(y) = -qN_A d_{max} = -\sqrt{2\varepsilon_s q N_A[V(y) + 2\Phi_B]} \tag{2.16}$$

Eq. 2.14 can also be written as

$$Q_n(y) = -C_{ox}[V_g' - 2\Phi_B - V(y)] + \sqrt{2\varepsilon_s q N_A[V(y) + 2\Phi_B]} \tag{2.17}$$

The voltage drop across an element, dy, along the channel is given by

$$dV = I_d dR = I_d dy/[W\mu_n|Q_n(y)|] \tag{2.18}$$

Using the boundary condition of $V = 0$ at $y = 0$ and $V = V_d$ at $y = L$, and Eqs. (2.17) and (2.18), the following I-V characteristics can be obtained

$$I_d = \frac{W}{L} \mu_n C_{ox}$$

$$\left\{ \left(V_g' - 2\Phi_B - \frac{V_d}{2} \right) V_d - \frac{2\sqrt{2\varepsilon_s q N_A}}{3 C_{ox}}[(V_d + 2\Phi_B)^{3/2} - (2\Phi_B)^{3/2}] \right\} \tag{2.19}$$

Eq. (2.19) indicates that I_d increases linearly with V_d, then gradually levels off approaching saturation. The mobility μ_n in the above equation is not a constant, rather it is a function of the effective electric field as described in Sec. 2.1.1. The effective electric field (E_{eff}) is the electric field averaged over the electron distribution in the channel. It can also be expressed as

$$E_{eff} = (Q_B + Q_n/2)/\varepsilon_s$$
$$\approx C_{ox}[V_g - V_t]/(2\varepsilon_s) \tag{2.20}$$

Substituting Eq. (2.20) into Eq. (2.1), the following is obtained

$$\mu_n = \mu_o\{\varepsilon_s U_{crit}/ [C_{ox}(V_g - V_t)]\}^{UEXP} \tag{2.21}$$

where U_{crit} is $2E_{crit}$. It is obvious that μ_n decreases as V_g is increased. Mobility modelling for device and circuit simulation will be discussed in Chapter 3, and the effect of scaling on mobility in Chapter 6.

I-V curves with various V_gs are shown in Fig. 2.12, calculated based on Eq. 2.19 for $V_d < V_{dsat}$. The dashed line marks the locus of the saturation drain voltage (V_{dsat}), beyond which I_d reaches its maximum and remains constant. For $V_d < V_{dsat}$, Eq. (2.19) can be approximated by

$$I_d = \mu_n(W/L)C_{ox}[(V_g - V_t) - V_d/2]V_d \tag{2.22}$$

where V_t, the threshold voltage, is given by

$$V_t = V_{FB} + 2\Phi_B + \sqrt{2\varepsilon_s q N_A(2\Phi_B)}/C_{ox} \tag{2.23}$$

For small V_ds, Eq. (2.22) can be further reduced to

$$I_d = \mu_n(W/L)C_{ox}(V_g - V_t)V_d \quad \text{for } V_d \ll (V_g - V_t) \tag{2.24}$$

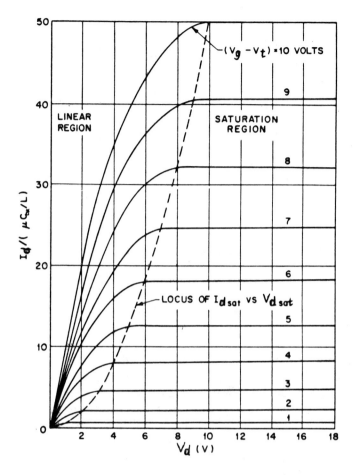

Figure 2.12   Idealized drain current vs. drain voltage of a MOSFET (Ref. 2).

This equation can then be rearranged as

$$I_d = \frac{WLC_{ox}(V_g - V_t)}{L/[\mu_n(V_d/L)]} \quad \text{for } V_d \ll (V_g - V_t) \tag{2.25}$$

where the numerator represents the total charge stored in the MOS capacitor and the denominator corresponds to the charge transit time, which is equal to the channel length divided by the carrier velocity v_n, where $v_n = \mu_n E = \mu_n V_d/L$. The MOSFET is now operated in the linear region and acts like a linear resistor. From Eq. 2.24, the threshold voltage can be approximated

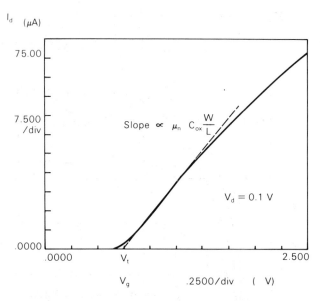

Figure 2.13 A typical experimental result of drain current (I_d) vs. gate voltage (V_g) at a drain voltage (V_d) of 0.1 V.

as the intercept of the I_d vs. V_g plot for a small V_d (e.g., 0.1 V). A typical experimental result of an I_d–V_g relation is shown in Fig. 2.13. The threshold voltage (V_t) is commonly measured by extrapolating the linear portion of the curve to the horizontal axis and the channel mobility (μ_n) is obtained from the slope of the dashed line. Notice a small but finite amount of current exists for V_gs which are slightly less V_t, due to weak inversion. At strong inversion, i.e., $V_g \geq V_t$, current is approximately linearly proportional to V_g. At a large V_d, the MOSFET is in a saturation region and Eq. 2.22 does not apply. Another way of obtaining threshold voltage is to define V_t as a gate voltage at which a significant current (normally 1 μA) is observed. This definition is obviously less rigorous but offers a convenient method to get a rough estimate.

The channel conductance g_d and the transconductance g_m are given as

$$g_d = (\partial I_d/\partial V_d)V_{g=\text{const}} = (W/L)\mu_n C_{ox}(V_g - V_t) \qquad (2.26)$$

$$g_m = (\partial I_d/\partial V_g)V_{d=\text{const}} = (W/L)\mu_n C_{ox}V_d \qquad (2.27)$$

For large V_ds, I_d saturates due to the pinchoff effect described earlier. From the locus of the saturation voltage in Fig. 2.12, V_{dsat} is approximately

$$V_{dsat} \cong V_g - V_t \qquad (2.28)$$

A more rigorous derivation can lead to similar approximation[1] by setting $Q_n(L) = 0$ at $V(L) = V_d = V_{dsat}$ in Eq. (2.17). Substituting Eq. (2.28) into

Eq. (2.22) yields the value of saturation current as

$$I_{d\text{sat}} \cong [W/(2L)]\mu_n C_{ox}(V_g - V_t)^2 \qquad (2.29)$$

The transconductance in the saturation region is then

$$g_m = (\partial I_d/\partial V_g)V_{d=\text{const}} = (W/L)\mu_n C_{ox}(V_g - V_t) \qquad (2.30)$$

In the above discussion, the series resistances associated with source and drain, were not considered. These series resistances impose ohmic drops between source and drain contacts and the channel. In the linear region, the effect of the series resistances can be realized by noting that[1]

$$1/g_d(\text{mea}) = 1/g_d + R_s + R_d \qquad (2.31)$$

where $g_d(\text{mea})$ is the measured channel conductance, g_d is the intrinsic channel conductance, and R_s and R_d are the source and drain series resistance, respectively. In the saturation region, the effective gate voltage is the applied gate voltage minus the source voltage, i.e. $V_g = V_g(\text{app}) - V_s$. Thus, the measured transconductance is

$$1/g_m(\text{mea}) = dV_g(\text{app})/dI_d = 1/g_m + R_s \qquad (2.32)$$

The drain resistance R_d does not appear in Eq. (2.32) because the change of V_d caused by IR drop has no effect on saturation current. The saturation voltage $V_{d\text{sat}}$ will, however be increased. The transconductance in the linear region is now more complicated because current depends on both gate voltage and drain voltage as follows

$$I_d = (W/L)\mu_n C_{ox}[V_g(\text{app}) - I_d R_s - V_t][V_d(\text{app}) - I_d(R_s + R_d)] \qquad (2.33)$$

The mathematics of deriving measured transconductance in the linear region is complicated and can be found elsewhere.[8] However, note that I_d and g_m in the linear region are degraded by both R_s and R_d.

Subthreshold region.   A previous section described the operation and basic characteristics of a MOSFET when its gate voltage exceeded the threshold voltage and the device was completely turned on. Another mode of operation is the weak inversion mode in which the Si surface has changed to n-type, but is not strongly inverted, meaning minority carrier concentration is still lower than the bulk doping concentration. Unlike in the linear region in which the onset of strong inversion occurs and the minority charge density increases linearly with V_g, in the subthreshold mode, minority carrier concentration is too low to change the band bending significantly, thereby resulting in a low electrical field along the channel. Both the low carrier concentration and the low field region make the drift current much less than the diffusion current, hence subthreshold current is dominated by the diffusion component and is given as[9]

$$I_d = AqD_n dn(y)/dy \cong Wd_{inv}(qD_n n_s)/L \qquad (2.34)$$

where A is the cross-sectional area of the electron flow, d_{inv} is the inversion layer thickness, D_n is the electron diffusion constant, and $n(y)$ and n_s are the minority carrier densities at Point y and at the source, respectively. Fig. 2.14 is a cross-sectional schematic that illustrates the notation used in Eq. 2.34. The minority carrier density at the drain is negligible for drain voltages larger than a few kT/q (e.g., $V_d > 0.1$ V at room temperature). The carrier density (n_s) is equal to $(n_i^2/N_A)e^{(q\Phi_s/kT)}$, and Φ_s is the surface potential at the source. Since n_s is exponentially dependent on the channel potential, the majority of the inversion charges are within the distance at which the potential drops by kT/q. Therefore, the inversion thickness (d_{inv}) is in the range of $kT/(qE_s)$ where E_s is the weak-inversion surface field given by $\sqrt{2qN_A\Phi_s/\varepsilon_s}$. Thus, I_d can be approximated as

$$I_d \approx qD_n(W/L)\,(kT/q)(qN_A\Phi_s/\varepsilon_{si})^{-1/2}(n_i^2/N_A)e^{(q\Phi_s/kT)} \qquad (2.35)$$

Eq. 2.35 indicates that I_d is independent of V_d. However, for short-channel devices, higher V_d causes Φ_s to increase, giving larger I_d. This change is due to the increase of V_d and it causes an expansion of the depletion layer associated with the reversely-biased drain-substrate junction. This depletion encroachment tends to lower the potential barrier, which prevents electrons from diffusing from source to drain. As a result, additional current is ob-

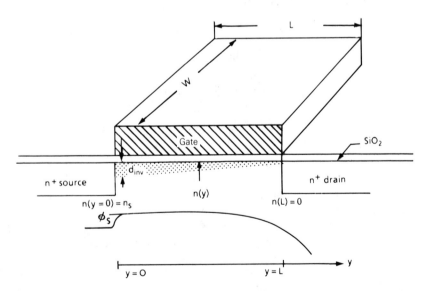

Figure 2.14   A cross-sectional schematic of a transistor illustrating subthreshold conduction.

served at zero or a few-tenths volt V_g. This V_d induced current is often referred to as punchthrough current because it is due to the drain depletion region punching through the channel. This punchthrough phenomenon is explained in Chapters 3 and 6.

Because $\mathbf{\Phi}_s$ is roughly proportional to V_g in the subthreshold region,[2] I_d is also exponentially dependent on V_g. An important device parameter in the subthreshold region is the slope of the log I_d vs. V_g. The inverse of the slope is defined as the subthreshold swing (S_t)

$$S_t = \ln10(d\ln I_d/dV_g)^{-1} = 2.3(kT/q)(1 + C_d/C_{ox})$$
$$= 2.3\,(kT/q)\,[1 + \varepsilon_{si}t_{ox}/(\varepsilon_{ox}d)] \tag{2.36}$$

S_t is defined in units of mV per decade. Ideally, an abrupt change in current should occur as the voltage varies; this change would indicate a small S_t. If the S_t is small, current can be turned off effectively by reducing V_g. For example, at an S_t value of 100 mV/dec, current drops from 1 μA to 1 pA for a 0.6 V decrease in V_g. If V_t is > 0.6 V, where V_t roughly corresponds to 1 μA current, then at $V_g = 0$ V, current is less than 1 pA.

On the other hand, a large S_t (meaning a gradual slope in log $I - V_g$) often results in a significant amount of current at the off state. From Eq. (2.36), the value of an S_t can be made smaller by using a thinner oxide (t_{ox}) or a lower doping concentration for a larger depletion width (d). Also, the value of S_t increases linearly with temperature. Typical experimental subthreshold characteristics are shown in Fig. 2.15 for various substrate bias voltages.

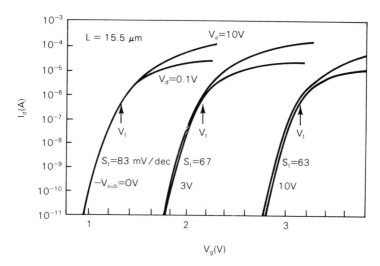

Figure 2.15 Experimental subthreshold characteristics for a long-channel MOS-FET, S_t is the subthreshold swing and V_{sub} is the substrate bias (Ref. 2).

Because the depletion width increases when a substrate bias is applied, the subthreshold swing decreases according to Eq. (2.36). Meanwhile, the threshold voltage increases. In an n-channel case, V_{sub} is normally negative with respect to the source voltage. This substrate effect (also referred to as body effect) can be understood by examining the threshold expression in Eq. 2.23. With a substrate bias (V_{sub}), the depletion width in the semiconductor is enlarged and $2\Phi_B$ in the bulk charge term is replaced by ($2\Phi_B + |V_{sub}|$) as

$$V_t = V_{FB} + 2\Phi_B + \sqrt{2\varepsilon_{si}qN_A(2\Phi_B + |V_{sub}|)}/C_{ox} \qquad (2.37)$$

A threshold increase due to a substrate bias can be written as

$$\Delta V_t = \gamma\left[\sqrt{(2\Phi_B + |V_{sub}|)} - \sqrt{2\Phi_B}\right] \qquad (2.38)$$

where γ defined as the following body effect coefficient:

$$\gamma = \sqrt{2\varepsilon_{si}qN_A}/C_{ox} = \sqrt{2\varepsilon_{si}qN_A}\, t_{ox}/\varepsilon_{ox} \qquad (2.39)$$

Fig. 2.16 shows delta V_t as a function of $\sqrt{V_{sub}}$ with different values of γ. Eq. 2.39 shows that γ can be reduced with lower doping concentration and/or thinner oxide.

The significance of the body effect is the change of the threshold voltage and the decrease of drain current at a given gate voltage. In an NMOS circuit, this effect often occurs due to the increase of an FET source voltage during circuit operation. As a result, V_{sub} is negative with respect to the source voltage, thereby causing V_t to increase and I_d to decrease. Consequently, the circuit runs at a lower speed and may not even function properly.

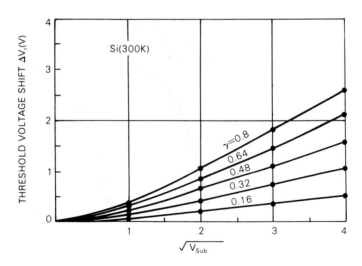

Figure 2.16   Threshold voltage shift vs. substrate reverse bias for various γ values (Ref. 2).

The discussion of an inverter circuit in Chapter 4 will illustrate the significance of the body effect on circuit performance.

MOSFET types.   All derivations previously mentioned use n-channel transistors as an example.   Similar results can be obtained for p-channel devices with reverse polarity for all terminal voltages and threshold voltage. Table 2.1 summarizes important equations for n- and p-channel MOSFETs. In both cases, the devices are off unless a sufficient gate voltage is applied. This type of device (normally-off) is called an enhancement mode device. Another type of transistor is normally on unless reverse-polarity V_g (e.g., negative V_g for an n-channel) is applied to turn it off.   These devices (normally-on) are called depletion mode devices.

For an n-channel depletion device, the threshold voltage is made negative by adding an arsenic ion implantation during device fabrication.   A total of four different types of MOSFETs result, as shown in Fig. 2.17.   The corresponding I-V characteristics are also shown.   In n-channel cases, a positive V_g must be applied to turn on an enhancement-mode device.   But, a negative V_g must be applied to turn a depletion-mode device off.   A depletion-mode device delivers current at $V_g = 0$ and is often used as the load device in NMOS logic circuits.   In p-channel cases, one applies a negative V_g to turn on an enhancement-mode device, but a positive V_g to turn off a depletion-mode device.   However, depletion-mode p-channel devices are normally not used in present MOS technologies.   In a CMOS circuit, enhancement mode p-channel MOSFETs are often used as load devices.   A depletion-mode device is commonly made by forming a thin n-type surface layer in the p-substrate for n-channel cases or a thin p-type surface layer in the n-substrate for p-channel cases.   Thus, source and drain are connected with the same type of semiconductor layer unless this layer is fully depleted by applying the appropriate gate voltage.   The presence of this surface layer however changes normal MOSFET operation from surface-channel conduction to buried-channel conduction.

2.4 BURIED CHANNEL DEVICES

Discussions so far have covered device operations in surface-channel MOSFETs in which carriers propagate at the semiconductor surface.   However, in some devices such as depletion-mode devices, carriers propagate slightly under the semiconductor surface.   This type of device is called a buried-channel device.   Fig. 2.18 shows the cross-section of a buried-channel nMOS device.   Notice the n-type layer with the doping concentration N_D and junction depth x_j.   The device has two depletion regions.   The bottom region originated from the p-n junction, and the top region is associated with the MOS capacitor.   The depletion width is not uniform because, under the bias

TABLE 2.1 Important Equations for n- and p-Channel MOSFETs.

	N-Channel $V_d > 0,\ V_g > 0,\ V_T > 0,\ V_{sub} \leq 0,\ V_s = 0$	P-Channel $V_d < 0,\ V_g < 0,\ V_T < 0,\ V_{sub} \geq 0,\ V_s = 0$
Threshold voltage	$V_t = \Phi_{ms} - \dfrac{Q_{fc}}{C_{ox}} + 2\Phi_B + \dfrac{\sqrt{2\epsilon_s q N_A(2\Phi_B + \lvert V_{sub}\rvert)}}{C_{ox}}$ Where $\Phi_B = \dfrac{kT}{q}\ln\left(\dfrac{N_A}{n_i}\right)$	$V_t = \Phi_{ms} - \dfrac{Q_{fc}}{C_{ox}} - 2\Phi_B - \dfrac{\sqrt{2\epsilon_s q N_A(2\Phi_B + V_{sub})}}{C_{ox}}$ Where $\Phi_B = \dfrac{kT}{q}\ln\left(\dfrac{N_D}{n_i}\right)$
$I - V$ in linear region	$I_d \simeq \dfrac{W}{L}\mu_n C_{ox}(V_g - V_t)\,V_d$ for $V_d \ll V_{dsat}$	$I_d \simeq \dfrac{W}{L}\mu_p C_{ox}\lvert V_g - V_t\rvert V_d$ for $\lvert V_d\rvert \ll \lvert V_{dsat}\rvert$
$I - V$ in saturation region	$I_d = I_{dsat} = \dfrac{W}{L}\mu_n C_{ox}\dfrac{(V_g - V_t)^2}{2}$	$I_d = I_{dsat} = \dfrac{W}{L}\mu_p C_{ox}\dfrac{(V_g - V_t)^2}{2}$
Saturation voltage	$V_{dsat} \simeq V_g - V_t$	$V_{dsat} \simeq V_g - V_t$
Subthreshold swing	$S_t = \left[\dfrac{d(\log I_d)}{dV_g}\right]^{-1} = 2.3\dfrac{kT}{q}\left(1 + \dfrac{C_d}{C_{ox}}\right)$	$S_t = \left[\dfrac{d(\log\lvert I_d\rvert)}{d\lvert V_g\rvert}\right]^{-1} = 2.3\dfrac{kT}{q}\left(1 + \dfrac{C_d}{C_{ox}}\right)$

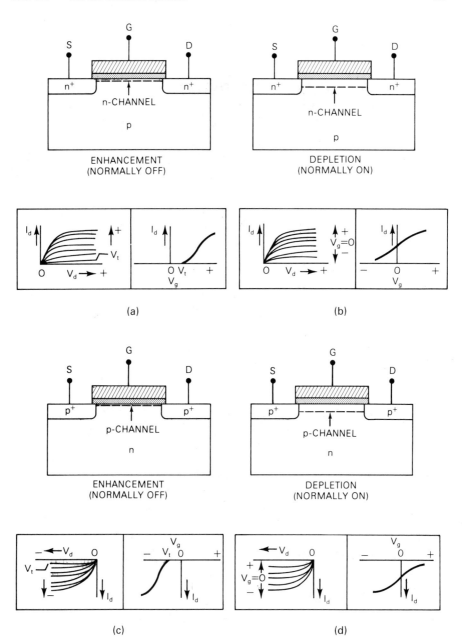

Figure 2.17 Four different MOSFET types and I-V characteristics: (a) enhancement *n*MOS; (b) depletion *n*MOS; (c) enhancement *p*MOS, and (d) depletion *p*MOS (Ref. 2).

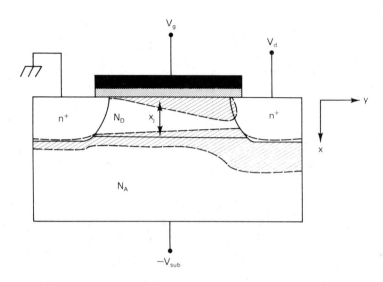

Figure 2.18   A typical buried-channel nMOS; the cross-hatched areas represent depletion regions (Ref. 2).

of drain voltage, the potential along the channel is different.   When the two depletion regions near the drain meet, the device is pinched off, thus causing the device to saturate.   This situation is similar to a Junction FET (JFET) operation.

A buried-channel MOSFET can be an enhancement-mode or a depletion-mode device, depending on the doping level and the depth of the thin surface layer.   A cross-sectional view through the region near the source of the FET is shown in Fig. 2.19a.   The cross-hatched areas correspond to depletion regions.   The corresponding electrical field, channel potential and free mobile carriers (electrons in this case) are also shown in the figure.   For a depletion mode buried-channel FET, the undepleted n-type surface layer decreases as the gate voltage becomes more and more negative and eventually the n-layer is completely depleted at a sufficiently negative gate voltage.   This voltage is the device's threshold voltage and can be obtained[2,10] by solving Poisson's equation

$$V_t - V_{FB} = -qN_D x_j (1/C_{ox} + 1/2C_D)$$
$$+ \sqrt{2\varepsilon_{si}qN_A(V_{bi} + V_{sub})}\ (1/C_{ox} + 1/C_D) \qquad (2.40)$$

where V_{bi} is the built-in voltage of the p-n junction and C_D is equal to ε_{si}/x_j. Notice that V_t is a function of N_D, x_j and t_{ox}.   If N_D and/or x_j is very large, as the magnitude of V_g increases, surface inversion occurs before the n-layer is completely depleted.   In fact, this instance occurs if the maximum depletion width (d_{max}) is smaller than the n-layer thickness (x_j), i.e.,

$$d_{max} = \sqrt{2\varepsilon_{si}(2\phi_B)/(qN_D)} = \sqrt{4\varepsilon_{si}kT\ln(N_D/n_i)/(q^2 N_D)} < x_j \qquad (2.41)$$

Under this condition, the depletion-mode device cannot be turned off regardless of the negative voltage applied on the gate.

For an enhancement-mode, buried-channel FET, the n-layer is completely depleted at zero gate voltage. A positive gate voltage must be applied such that the top depletion width (x_1 in Fig. 2.19a) is reduced to ($x_j - x_2$). Further increases in gate voltage would then open an undepleted channel for the current to conduct. This positive gate voltage is the threshold voltage and can be expressed as[11]

$$V_t = V_{FB} - qN_D (x_j - x_1)/C_{ox} - qN_D (x_j - x_1)^2/(2\varepsilon_{si}) \qquad (2.42)$$

The buried-channel MOSFET shown in Fig. 2.19, has two depletion capacitances in addition to the oxide capacitance. The potential minimum ($\mathbf{\Phi}_{min}$), and hence the free carriers are in the bulk Si rather than at the Si–SiO$_2$ interface as in the case of a surface-channel device. It is then expected that the carriers, because they are now farther away from the surface, are harder to modulate by gate voltage. The turn-off characteristic is normally less sharp in this type of device due to the larger subthreshold swing (S_t) which again can be given by the capacitive divider ratio as in Eq. 2.36

$$S_t = (2.3\ kT/q)(1 + C_2/C_1)$$

$$= (2.3\ kT/q)[1 + (\varepsilon_{ox}x_1 + \varepsilon_{si}t_{ox})/(\varepsilon_{ox}x_2 + \varepsilon_{ox}x_3)] \qquad (2.43)$$

where C_1 is the oxide capacitance(C_{ox}) in series with the depletion capacitance from the surface (ε_{si}/x_1) and C_2 is the depletion capacitance from the p-n junction. As shown in Fig. 2.19, $\mathbf{\Phi}_{min}$ is the potential at the position between these two capacitors. Compared to Eq. (2.36), C_1 in Eq. (2.43) is less than C_{ox} and C_2 ($C_2 = \varepsilon_{si}/(x_2 + x_3) \approx \varepsilon_{si}/x_3$) is normally comparable with C_d. Experimental results have shown that the subthreshold swing is worse in buried-channel devices especially when the channel length (L) is short (see Fig. 2.20).[12–14]

Buried-channel FETs, however, have some advantages. The corresponding mobility is higher because mobile carriers propagate away from the Si surface, thereby suffering less surface scattering. As a result, transconductance (g_m) can be higher even though $g_m \propto \mu_{ch}C_g$ and C_g, the effective gate capacitance, is somewhat reduced in a buried channel.

The short-channel effect, commonly expressed as V_t reduction due to channel shortening, is another concern in buried-channel devices. Threshold lowering measured in the saturation region ($V_d = 5$ V) always shows that buried-channel devices are more susceptible to the short-channel effect due to increased drain-induced barrier lowering.[15] Threshold roll-off in the linear region of a buried-channel could be better,[11] or worse,[13,14] depending on a particular device's design.

The behavior of an enhancement-mode buried-channel device is a strong function of gate voltage. At high V_g, i.e., $V_g \gg V_t$, the energy bands at the Si surface are lowered sufficiently to cause the device to operate in a surface-

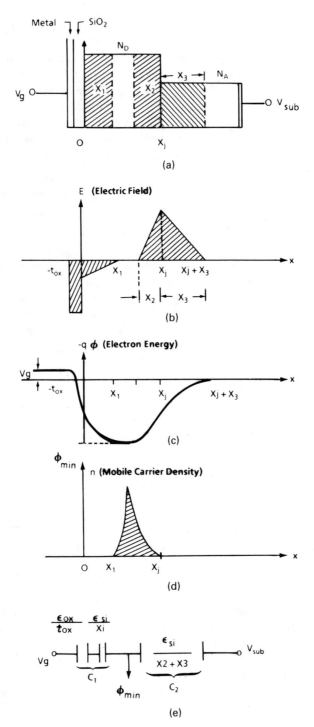

Figure 2.19 (a) Doping profile and depletion regions; (b) electric field; (c) electron potential; (d) mobile carriers; and (e) equivalent capacitances in sub-pinchoff condition of a buried-channel MOSFET.

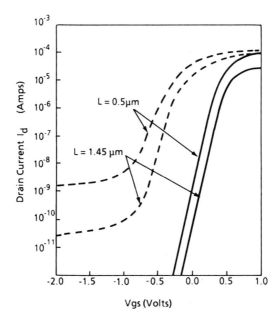

Figure 2.20 Experimental subthreshold characteristics of buried-channel (dashed lines) and surface-channel (solid lines) MOSFETs (Ref. 12, © 1981 IEEE).

channel mode.   Under this condition, the device no longer exhibits buried-channel characteristics such as high mobility.   At $V_g < V_t$, buried-channel behaviors such as subthreshold swing and V_t lowering, are preserved.   In CMOS, one type of FET is normally buried-channel due to a constraint in the standard fabrication process.   Chapter 6 contains more details on this constraint.

2.5 SHORT- AND NARROW-CHANNEL EFFECTS

When MOS transistors are made smaller (shorter or narrower) for a VLSI circuit, two-dimensional effects near the edges of the smaller transistor become significant.   In a short-channel MOSFET, the depletion c. rges near the ends of the channel are generated by the field lines from the source and drain.   The bulk charge (Q'_B) depleted by voltage on the gate is then significantly reduced because of the charge sharing phenomenon.   As shown in Fig. 2.21, the total charge Q'_B is now the charge in the cross-hatched trapezoid[16]

$$Q'_B = qN_AW_mW(L + L')/2$$
$$= qN_AW_mWL[1 - (\sqrt{1 + 2W_m/x_j} - 1)x_j/L] \tag{2.44}$$

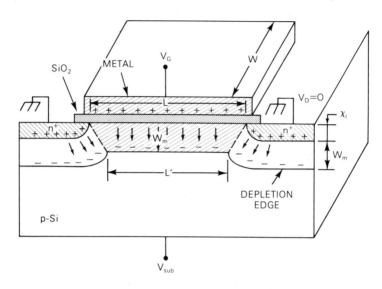

Figure 2.21 Charge sharing in a short-channel MOSFET (Ref. 2).

The threshold lowering is then

$$\Delta V_t = V_t(Q_B) - V_t(Q'_B)$$
$$= qN_A W_m x_j(\sqrt{1 + 2W_m/x_j} - 1)/(C_{ox}L) \qquad (2.45)$$

where W_m is the maximum depletion width, C_{ox} is the oxide capacitance per unit area, and x_j is the junction depth. Equation (2.44) is derived with both the source and the drain grounded. To minimize ΔV_t, a designer should use shallower source/drain junctions, thinner oxide thickness and increased channel length. Threshold lowering effect is more pronounced when a large drain voltage (V_d = 5 V) is applied because more depletion charge is contributed by the drain junction.

Another short-channel effect is the degradation of subthreshold characteristics. Figure 2.22 shows current-voltage characteristics in the subthreshold region for devices with various channel lengths. Both punchthrough current per unit channel width (I/W) and subthreshold swing (S_t) increase as the channel length (L) is decreased. The result is higher off-state leakage current which is undesirable for VLSI. Also, the subthreshold characteristics in a buried-channel device can be worse than that in a surface-channel device. These short-channel effects must be taken into account when designing CMOS transistors. These effects are discussed in Chapter 6.

The narrow-channel effect, though less critical, should also be considered. The threshold of a narrow-channel device is increased as the channel width is reduced. This threshold increase is due to the lateral extension of the depletion region along the channel width direction. The bulk charge,

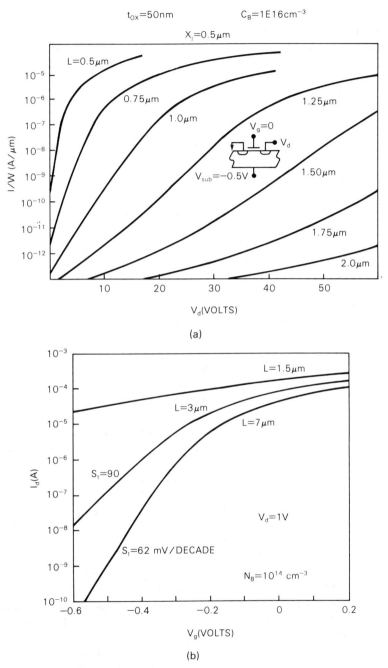

Figure 2.22   Subthreshold characteristics for MOSFETs with various channel lengths: (a) drain current vs. drain voltage at $V_g = 0$ V (Ref. 15, © 1979 IEEE); (b) drain current vs. gate voltage at $V_d = 1$ V (Ref. 2).

Q'_B term in the threshold expression, increases significantly as the channel width is decreased. In other words, one must now apply more gate voltage to invert the entire surface because of the field lines from the gate that terminate outside the channel.

REFERENCES

1. A. S. Grove, Physics and technology of semiconductor physics, John Wiley & Sons, p. 93, 1967.

2. S. M. Sze, Physics of Semiconductor Devices, 2nd ed., John Wiley & Sons, p. 46, 1981.

3. A. G. Sabnis and J. T. Clemens, "Characterization of the electron mobility in the inverted ⟨100⟩ Si surface," *IEDM Tech. Dig.*, p. 18, 1979.

4. S. C. Sun and J. D. Plummer, "Electron mobility in inversion and accumulation layers on thermally oxidized silicon surfaces," *IEEE Trans. on Electron Devices*, vol. ED-27, p. 1497, 1980.

5. D. Frohman-Bentchkowsky, "On the effect of mobility variation on MOS device characteristics," *Proc. IEEE*, vol. 56, p. 217, 1968.

6. C. A. Lee, R. A. Logan, R. L. Batdorf, J. J. Kleimack, and W. Wiegmann, "Ionization rates of holes and electrons in silicon," *Phys. Rev.*, vol. 134, A761, 1964.

7. C. R. Crowell and S. M. Sze, "Temperature dependence of avalanche multiplication in semiconductors," *Appl. Phys. Letts.*, vol. 9, p. 242, 1966.

8. P. I. Suciu and R. L. Johnston, "Experimental derivation of the source and drain resistance of MOS transistors," *IEEE Trans. on Electron Devices*, vol. ED-27, p. 1846, 1980.

9. J. R. Brews, "Subthreshold behavior of uniformly and nonuniformly doped long-channel MOSFET," *IEEE Trans. on Electron Devices*, ED-26, p. 1282, 1979.

10. T. E. Hendrickson, "A simplified model for subpinchoff conduction in depletion-mode IGFET's," *IEEE Trans. on Electron Devices*, ED-25, p. 435, 1978.

11. K. Nishiuchi, H. Oka, T. Nakamura, H. Ishikawa, and M. Shinoda, "A normally off type buried-channel MOSFET for VLSI circuits," in *IEDM*, p. 26, 1978.

12. T. N. Nguyen and J. D. Plummer, "Physical mechanisms responsible for short channel effects in MOS devices," in *IEDM Dig.*, p. 596, 1981.

13. T. N. Nguyen and J. D. Plummer, "A comparison of buried channel and surface channel MOSFET's for VLSI," presented at *40th Device Res. Conf.*, paper IIA-4, June, 1982.

14. G. J. Hu and R. H. Bruce, "Design tradeoff between surface and buried channel FET's," *IEEE Trans. Electron Devices*, ED-32, p. 584, 1985.

15. R. R. Troutman, "VLSI limitations for drain induced barrier lowering," *IEEE Trans. on Electron Devices*, ED-26, p. 461, 1979.

16. L. D. Yau, "A simple theory to predict the threshold voltage in short-channel IGFET's," *Solid-State Electronics*, vol. 17, p. 1059, 1974.

EXERCISES

1. In an n-type Si with an impurity concentration of 1×10^{17} cm^{-3}, calculate the impact ionization rate at an electrical field of 5×10^4 V/cm. If the Si is doped to p-type at 1×10^{17} cm^{-3}, calculate the impact ionization rate again at the same electrical field level.

2. Write a threshold expression for an ideal MOS capacitor, then modify it to account for oxide charges and work functions.

3. Assuming $V_{FB} = 0$, for a MOS capacitor with a 250 Å gate oxide and a 5×10^{16} cm^{-3} substrate doping concentration, calculate the threshold voltages for an n- and a p-channel MOS device.

4. Supposing that an oxide charge density of 5×10^{10} cm^{-2} exists in a MOS gate oxide close to the Si–SiO$_2$ interface and the MOS gate is made of aluminum with a work function of 4.3 V, calculate the threshold voltages for an n- and p-channel MOS device. The electron affinity for Si is 4.05 V.

5. For an n-channel MOSFET with 250 Å oxide, 1 μm channel length, 10 μm channel width, and 1 V threshold voltage, operated at $V_g = 5$ V and $V_d = 0.1$ V, calculate the total electron charges in the channel and the electron transit time, and show that the charge divided by the transit time is equal to the current calculated at the linear region.

6. Describe how the saturation voltage (V_{dsat}) is defined and derive $V_{dsat} = V_g - V_t$ as a first order approximation.

7. Both R_s and R_d are parasitic resistances. Discuss why R_s more importantly affects transistor performance in the saturation region. Also, derive the analytic expression for measured g_m in the linear region as a function of R_s and R_d.

8. If a circuit designer wants no more than 1 pA off-state leakage and also wants $V_t = 0.9$ V (V_t is defined as the gate voltage giving 1 μA current), what S_t value should you achieve?

9. Using Eq. (2.43), calculate the buried-channel S_t value for $x_1 = 500$ Å, $t_{ox} = 250$ Å, $x_2 \ll x_3 = 3000$ Å and compare it with the S_t value calculated for a surface-channel MOSFET with the same t_{ox} and a depletion width of $d = x_3$.

3

MOS MODELLING

This chapter describes the use of modelling tools for CMOS device design and process integration. The chapter will first describe important process modules, then review the process and device models available to date. The basic principles, assumptions, approximations and limitations of these models are discussed. Ways to use these models as design tools are introduced and examples are presented. The last section presents models that have been developed to extract MOSFET parameters for circuit simulation.

3.1 MODELLING FOR CMOS TECHNOLOGY DEVELOPMENT

A device technologist generally sits in the box located in the middle of Fig. 3.1. This person designs devices for circuit applications and develops processes to fabricate these devices. On one hand, he or she has to know a circuit's needs to satisfy customers; on the other hand, this person must also know process capabilities and constraints to establish a realistic IC process. The person should have knowledge of all available process modules and be able to integrate them to make devices that meet the customer's (who is often the circuit designer) specifications.

The goal of process integration is to build a process architecture rather than develop individual processes.   To perform these duties, process and device modelling are important because a trial-and-error approach by experimentation is expensive and time-consuming.   Also, an in-depth understanding of device operation often cannot be gained by performing experiments.   As illustrated in Fig. 3.1, modelling tools must be in place to support process simulation and device design.   It is necessary, however, to verify the modeling results through experimentation and testing using a set of test structures often designed with various dimensions.   Device parameters and design rules can then be generated and supplied to IC designers for circuit design and simulation.

The need for process and device modelling has become more concrete as the IC industry has moved into the VLSI era.   Operations in small-geometry devices can no longer be completely governed by straightforward, one-dimensional analytic solutions. The equations discussed in Chapter 2 are

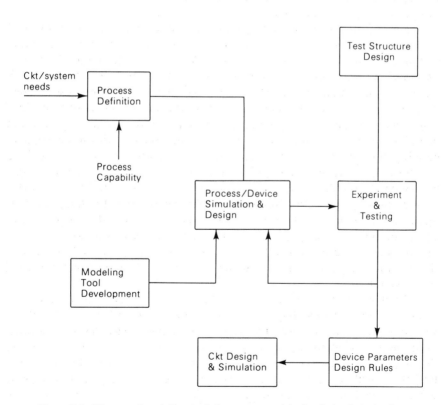

Figure 3.1   The use of modelling tools for process and device design in technology development.

necessary but not sufficient for designing VLSI devices. Complex two- or even three-dimensional effects in realistic situations however can be numerically simulated with reasonable accuracy. Meanwhile, IC processes for VLSI are becoming more complicated and requiring more time to complete. As a result, an empirical approach through experiment iteration is no longer a sensible way to establish a VLSI process.

Computer-aided modelling has become a powerful design tool because it allows many initial experiments to be performed through computer iterations instead of silicon cycling. Fewer experiments are needed to verify modelling results and fine-tune the final process.

Another benefit of modelling is a better understanding of device physics, which may not be obtainable otherwise. For example, device phenomena such as the high electrical field in short-channel transistors cannot be measured directly.

Lastly, stringent control in VLSI processes requires the understanding of the sensitivities of device characteristics on individual process variations. Through modelling, process-induced statistical fluctuations in device and circuit performance can be understood, then minimized.

Fig. 3.2 shows the modelling tools needed at all the different levels of developing a technology. The names within parentheses represent simulators and models currently available publicly. A bottom-up technique first specifies models for all individual process modules, then integrates them through the use of one-dimensional (1-D) and two-dimensional (2-D) process models. After the process parameters such as oxide thickness and impurity profiles have been obtained, these parameters are input into a device model, often two-dimensional, to gain device characteristics such as threshold voltage, mobility, body effect coefficient, etc. Process simulation and device modelling are closely coupled and exercised iteratively so that an adequate process can be obtained and the desired device parameters can be reached. Device characteristics are modified by comparing them with current-voltage relations measured from fabricated devices. This exercise is referred to as parameter extraction because the final device parameters are essentially extracted from real devices. These extracted device parameters are then used in a circuit simulator to analyze circuit performance. SPICE is the most commonly used circuit simulator.

Models for individual process modules have been described extensively in other books[1-4]; this chapter will only describe them briefly. Emphasis will be placed on models for process integration and device design. Because using these models for device design is the main objective, the physics and the assumptions that the models are based on will be outlined so that their limitations will be clear. The numerical techniques employed in developing these models are of interest to tool builders, but are not as important for model users and hence will not be discussed in detail.

Modeling Tools

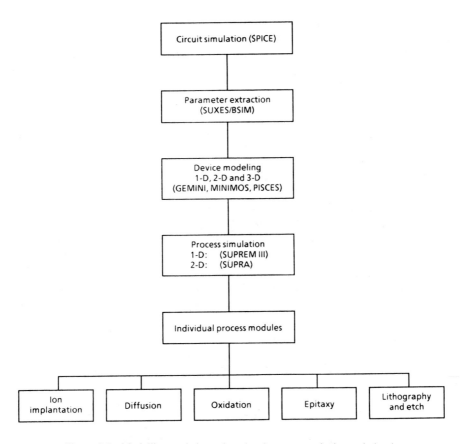

Figure 3.2 Modelling tools in various levels: process, device and circuit.

3.2 MODELS FOR INDIVIDUAL PROCESS MODULES

3.2.1 Ion Implantation

Ion implantation is a process by which ionized atoms are accelerated into a semiconductor substrate through high electrical field. The resultant profile, in first approximation, follows a Gaussian distribution

$$n(x) = n(R_p)\exp[-(x - R_p)^2/(2\sigma_p^2)] \qquad (3.1)$$

where the two moments R_p and σ_p are the peak (range) and the standard deviation (straggle) of the Gaussian distribution. The range and the straggle depend on the ion energy and the substrate material. The peak concentration, $n(R_p)$ is approximately $0.4 \times \text{dose}/\sigma_p$. Mathematically, $n(x)$ drops by two orders of magnitude at $3\,\sigma_p$, three orders of magnitude at $3.7\,\sigma_p$ and four orders of magnitude at $4.3\,\sigma_p$.

Experimental results from many ions such as boron and arsenic are found to be asymmetrical, hence a double half-Gaussian profile[1] including three moments provides a better approximation. The profile is

$$n(x) = n(R_p)\exp[-(x - R_p)^2/(2\sigma_{p1}^2)] \quad \text{for } x < R_p \qquad (3.2a)$$

$$n(x) = n(R_p)\exp[-(x - R_p)^2/(2\sigma_{p2}^2)] \quad \text{for } x > R_p \qquad (3.2b)$$

A more accurate analysis uses four moments. In addition to the range and the straggle, there are "skew" describing the asymmetry of the profile and "kurtosis" corresponding to the tail of the profile. This analysis leads to Pearson IV distribution[5] which fits experimental data much better than a symmetrical Gaussian distribution.

Fig. 3.3 shows the comparison for boron atoms implanted into amorphous silicon.[5] Implanting to an amorphous substrate avoids the channeling effect that produces an extended tail in the profile. The first two moments for the most commonly-used impurities in silicon substrates are shown in Fig. 3.4.[6] The others can be obtained through lookup tables[1] and the use of interpolation.

In a typical IC process, ions are implanted into selected areas by using masking materials such as a layer of polysilicon, oxide, nitride, or photoresist. The masking material with the proper thickness blocks ions, hence protecting

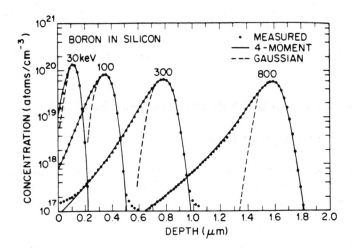

Figure 3.3 The distributions of boron atoms implanted into amorphous Si at different implant energies without annealing (Ref. 5).

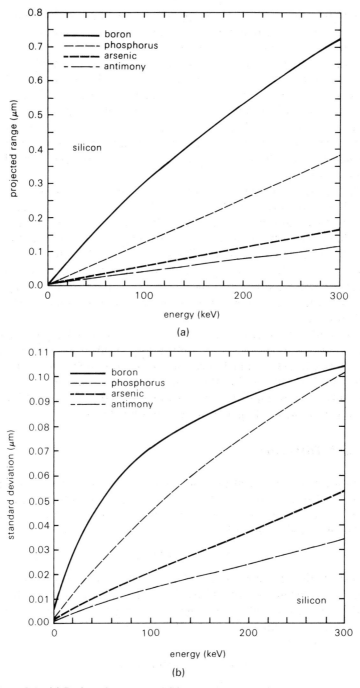

Figure 3.4   (a) Projected ranges; and (b) standard deviations of boron, phosphorus, arsenic and antimony (Ref. 6).

the underlying substrate. The "stopping powers" of different masking materials for commonly used impurities are illustrated by the ranges and straggles for various implant energies summarized in Table 3.1.[7] Interpolation and extrapolation can be used for other energies. Another option fits simple polynomials $(a_i E^i)$ to the $R_p(E)$ and σ_p (E) data shown in Fig. 3.4 for Si. Once the coefficients $(a_i$'s) of the polynomial are extracted, R_p and σ_p can be readily calculated for any given implant energy. Depending on the application, the thickness of a masking layer in general needs to be R_p + (3–4) σ_p. This method however does not account for the area near the mask edge which is not fully protected due to the lateral straggle caused by ion scattering in the substrate.

This scattering effect is important because the source and drain of a MOSFET is commonly formed using a polysilicon gate as a self-aligned mask. The lateral extension of the source and drain results in a shortening of the electric channel length and an increase of the gate-to-drain and gate-to-source overlap capacitance. The lateral straggle has been modelled by a complementary error function in the proximity of the mask edge.[8] For a silicon substrate, the lateral straggle for boron is similar to its vertical straggle, but the lateral straggle for phosphorus or arsenic is approximately 80% of its vertical straggle.[4]

The implanted species must be activated through a high temperature anneal. During this process, impurities move to substitutional sites and become electrically active and lattice damage is also removed. Meanwhile, impurity atoms diffuse during the high temperature anneal. The diffusion process causes the as-implanted impurity profile to be more relaxed, resulting in a final profile with a larger straggle and a longer tail.

3.2.2 Diffusion

Diffusion is a physical phenomenon where particles such as impurity atoms in a semiconductor tend to flow from high concentration to low concentration. Diffusion models are based on modified Fick's laws. The first Fick's law in one-dimension is

$$F(x) = -D\partial n(x)/\partial x \qquad (3.3)$$

When $F(x)$ is the diffusion flux, $n(x)$ is the impurity concentration, and the proportional constant D is the diffusivity. The second Fick's law is basically derived by using the continuity equation stating that particle generation is equal to the divergence of particle flux. This law leads to the following diffusion equation:

$$\partial n/\partial t = D(\partial^2 n/\partial x^2) \qquad (3.4)$$

and D depends on temperature as

$$D = D_o \exp(-E_a/kT) \qquad (3.5)$$

TABLE 3.1 Projected Ranges (Rp) and Standard Deviations (σ_p) of As, B, and P Ions Implanted into Four Different Substrates with Various Energies (Ref. 7)

Ion	As							
Sub.	Si		SiO_2		Si_3N_4		AZ 111	
Energy (keV)	Rp (A)	σ_p (A)	Rp (A)	σ_p (A)	Rp (A)	σ_p (A)	Rp (A)	σ_p (A)
10	97	36	77	26	60	20	412	78
50	322	118	260	85	202	66	1343	244
100	582	207	473	151	367	118	2375	415
150	845	292	690	214	534	167	3397	575
200	1114	374	913	275	706	214	4430	727
260	1445	470	1187	347	918	270	5688	904
300	1671	533	1374	394	1063	307	6539	1018
360	2015	624	1659	462	1283	360	7829	1183
400	2247	684	1851	506	1431	394	8698	1290

Ion	B							
Sub.	Si		SiO_2		Si_3N_4		AZ 111	
Energy (keV)	Rp (A)	σ_p (A)	Rp (A)	σ_p (A)	Rp (A)	σ_p (A)	Rp (A)	σ_p (A)
10	333	171	298	143	230	111	1086	270
50	1608	504	1606	483	1239	377	5683	868
100	2994	710	3104	710	2396	555	10564	1202
150	4205	834	4434	851	3424	666	14738	1391
200	5297	921	5643	951	4358	744	18445	1518
260	6496	999	6977	1040	5390	815	22476	1628
300	7245	1040	7812	1087	6037	852	24976	1685
360	8309	1092	9001	1147	6957	900	28515	1757
400	8987	1121	9759	1181	7544	926	30759	1797

Ion	P							
Sub.	Si		SiO_2		Si_3N_4		AZ 111	
Energy (keV)	Rp (A)	σ_p (A)	Rp (A)	σ_p (A)	Rp (A)	σ_p (A)	Rp (A)	σ_p (A)
10	139	69	108	48	84	37	480	114
50	607	256	486	185	376	143	2060	427
100	1238	456	1002	333	774	259	4182	765
150	1888	628	1537	461	1188	358	6365	1052
200	2539	775	2073	571	1602	444	8544	1296
260	3309	928	2709	685	2094	533	11110	1543
300	3812	1017	3125	751	2415	584	12783	1685
360	4549	1136	3735	839	2887	653	15227	1871
400	5029	1206	4133	891	3194	694	16813	1980

where E_a is the activation energy which is roughly 3–4 times of the silicon energy gap for both donors and acceptors in silicon.[3] The diffusivity is proportional to mobility through the well-known Einstein's relationship: $D = \mu kT/q$. Solving Eq. (3.4) with the boundary conditions $\partial n/\partial x|_{(0t)} = 0$ and $n(\infty,t) = 0$, and the initial pre-deposited profile approximated by a delta function with the total dose of impurities in the substrate equal to Q, the diffusion profile is

$$n(x, t) = [Q/\sqrt{\pi Dt}]\exp(-x^2/4Dt). \tag{3.6}$$

Again, the profile is a Gaussian distribution with the peak at the Si surface and the standard deviation at $\sqrt{2Dt}$ below the surface.

This simple theory holds true for intrinsic Si in which impurities exist at a concentration less than $n_i(T)$, the Si intrinsic concentration at the process temperature. $n_i(T)$ varies from 5 to 9×10^{18} cm^{-3} as the process temperature is changed from 900 to 1100°C. The above model fails when Si becomes extrinsic. One explanation is that ionized donors or acceptors result in a built-in electric field that can affect the motion of charged diffusant. However, attempts to include the electric field effect for explaining observed diffusivity enhancement in extrinsic Si[9] have not been fully successful.[10] Diffusivity enhancement is presently believed to be due to the interaction with lattice point defects such as vacancies and interstitials. Simplified forms for boron and arsenic in Si have been obtained from computer simulations and are[10]

$$D = D_i(1+3n_i/n)/4 \qquad \text{for boron} \tag{3.7}$$

$$D = D_i(1+100n/n_i)/101 \qquad \text{for arsenic} \tag{3.8}$$

where D_i is the intrinsic diffusivity and n is the carrier concentration. The situation is more involved for phosphorus because its diffusivity strongly depends on doping concentration due to vacancy enhancement. D increases proportionally to $(n_i/n)^2$ until the concentration reaches to $3 - 4 \times 10^{20}$ cm^{-3} due to Si band gap narrowing.[10] Intrinsic (doping less than 10^{18-19} cm^{-3}) diffusivities for common impurities in Si have been obtained and are shown in Fig. 3.5.[6]

Another important consideration for process modelling is the effect of oxidation enhanced diffusion. Due to the enhancement of point defects during oxidation, enhanced diffusion is observed for all commonly used impurities in Si. For example, phosphorus diffusion is enhanced by a factor of 1.8 in dry oxidation and 3.3 in wet oxidation.[10,11]

Since diffusion is basically an isotropic phenomenon, the lateral diffusion is essentially identical to its vertical diffusion. The diffusion equation in two-dimensional form is

$$\partial n/\partial t = D(\partial^2 n/\partial x^2 + \partial^2 n/\partial y^2) \tag{3.9}$$

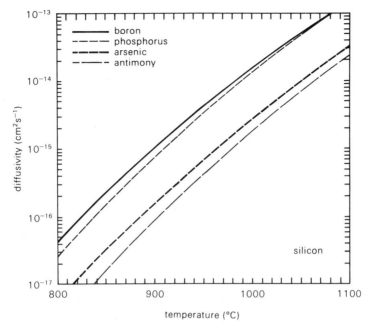

Figure 3.5 Intrinsic diffusion coefficient of boron, phosphorus, arsenic and antimony in Si versus temperature (Ref. 6).

The corresponding solution is

$$n(x,y,t) = [Q/(\pi Dt)]\exp(-x^2/4Dt)\exp(-y^2/4Dt) \qquad (3.10)$$

Diffusion occurs for any non-uniform impurity doping profile at high ($>800°C$) temperatures. As a result, impurity atoms, whether originally introduced by implantation, deposition, or epitaxy, diffuse during any high temperature steps in an IC process. The diffusion rate primarily depends on Dt product, and it depends secondly on doping concentration and oxidation.

3.2.3 Oxidation

Oxidation is a process in which a layer of silicon dioxide is thermally grown on a silicon wafer. At an elevated temperature, a gas containing oxygen or water vapor flows through a furnace and onto a silicon surface causing a chemical reaction. In the growth of an oxide, the thickness of the silicon layer that is consumed is 45% of the total oxide thickness. The oxidation rate is described by the following well-known expression[11]:

$$t_{ox}^2 + At_{ox} = B(t+\tau) \qquad (3.11)$$

where t_{ox} is the oxide thickness, t is time, τ is related to the initial oxide thickness by $\tau = (t_{oxi}^2 + At_{oxi})/B$, and A and B correspond to the following linear and parabolic growth rates:

$$A = P_{O2}K_p/K_l \qquad\qquad (3.12a)$$

$$B = P_{O2}K_p \qquad\qquad (3.12b)$$

where K_l and K_p are the linear and parabolic growth coefficients and P_{O2} is the normalized partial pressure of O_2. In general, K_l and K_p are functions of silicon crystal orientation, oxidizing ambient, and temperature. Both the values of K_l and K_p are larger for wet oxidation containing H_2O and smaller for dry oxidation with O_2. Each depends exponentially on temperature through a single activation energy.[12] The solution of Eq. (3.11) is plotted in Fig. 3.6 and compares well with a large number of experimental results obtained under various conditions.[12]

Impurity segregation during oxidation is crucial in changing impurity profiles in silicon. Impurity redistribution during oxidation is governed by impurity diffusivity in oxide and the moving rate of the silicon/oxide boundary.

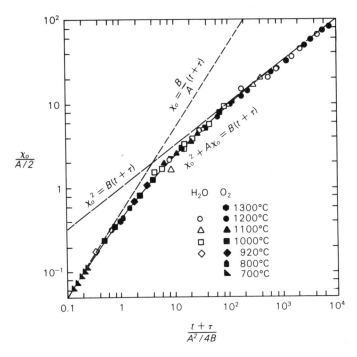

Figure 3.6 The general relationship for Si oxidation and its two limiting forms. A and B are the linear and parabolic constants and $x_o = t_{ox}$, the oxide thickness (Ref. 3).

In equilibrium, the ratio of the impurity concentration in silicon to that in oxide is a constant and can be defined as the segregation coefficient, m.   For boron in silicon, the effective m values are[13]

$$m = 13.4 \exp(-0.33\text{eV/kT}) \qquad \text{in dry oxidation} \qquad (3.13a)$$

$$m = 104 \exp(-0.66\text{eV/kT}) \qquad \text{in wet oxidation} \qquad (3.13b)$$

For wet oxidation, m decreases by a factor of 0.6 if the silicon orientation is altered from (100) to (111). m is orientation-independent for dry oxidation. m values of boron are less than one under ordinary oxidation temperatures. The result is a depletion of boron concentration at the silicon surface due to the oxide consuming boron atoms.   For phosphorus, m values are greater than one (quoted around 10) implying a pile-up of phosphorus at the silicon surface.[14]   The exact solution for the impurity distribution in silicon and oxide is complicated, but the impurity concentration at the silicon surface (C_s) has been obtained[3] and is a function of several parameters:

$$C_s = C_s(m, \, D_{ox}/D_{si}, \, B/D_{si}) \qquad (3.14)$$

where D_{ox} and D_{si} are the impurity diffusivities in oxide and silicon, respectively, and B is the parabolic oxidation rate.   Because these parameters depend on oxidation conditions, it is expected that C_s is a function of the oxidation temperature and ambient as shown in Fig. 3.7a and 3.7b for boron and phosphorus.   Increasing the oxidation rate by using wet rather than dry ambient results in more impurity segregation.   Also, lower temperatures enhance impurity segregation.

3.2.4 Epitaxy

Epitaxy is the growth of a single crystal semiconductor film on a single crystal substrate of the same semiconductor.   Techniques of growing semi-conductor layers on substrates of different types have also been recently developed, and are referred to as hetero-epitaxy.   Conventional epitaxial growth is called homo-epitaxy.   This section will discuss the growth of single crystal silicon on a silicon substrate.   This process is important for bipolar or CMOS IC fabrication because it is the only way to form a lightly-doped layer on a heavily-doped substrate.   Other processes such as implantation or diffusion normally result in a heavily-doped layer on a lightly-doped substrate. The kinetics of epitaxial growth and the associated mass-transfer models are clearly discussed elsewhere.[3]   However, it is worthwhile to discuss impurity redistribution during epitaxy because it affects device characteristics such as latchup in CMOS.

An epitaxial film is normally grown at elevated temperatures between 1000 to 1200°C.   Although it is desirable to have a sharp gradient for the doping concentration between epi-film and substrate, impurity diffusion dur-

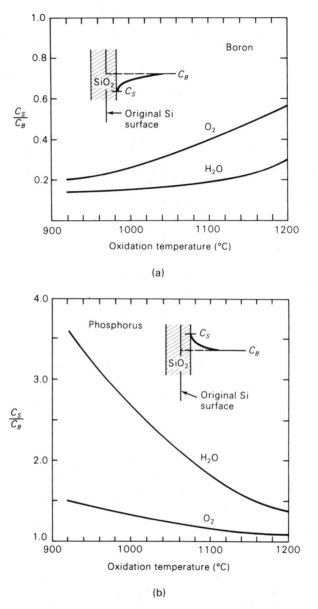

(a)

(b)

Figure 3.7   Normalized surface concentrations of: (a) boron; and (b) phosphorus in silicon after thermal oxidation. C_S and C_B are the surface and bulk concentrations, respectively (Ref. 3).

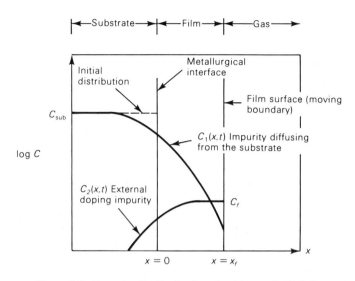

Figure 3.8   Impurity distribution in epitaxial growth (Ref. 3).

ing the high-temperature epi process always smooths out the doping profile at the epi/substrate interface.   Impurity atoms in the heavily-doped substrate diffuses into the epi film resulting in a smeared epi/substrate interface shown as C_1 in Fig. 3.8.[3]   Although the external doping impurity introduced during epi growth also diffuses from the epi film into the substrate, the effect is negligible because the substrate is already heavily doped.   The actual impurity redistribution is complicated because of the moving boundary during epi growth.   However, for cases where the epi growth rate is more than 5 times the impurity diffusion rate, impurity distributions can be approximated as complementary error functions by[15]

$$C_1(x,t) = \frac{1}{2}C_{sub}\,\text{erfc}[x/(2\sqrt{Dt})] \qquad (3.15)$$

where C_{sub} is the substrate concentration.   This approximation is reasonable for conventional epi growth process.   As an example, Fig. 3.9 shows the boron impurity distribution in a p-on-p^+ epi film grown for a CMOS application.[16]   Even though a 4 μm epi film is grown, the actual lightly-doped epi layer is reduced due to the out-diffusion of boron from the heavily-doped substrate into the lightly-doped epi layer.   The out-diffusion occurs during high-temperature epi growth, n-well diffusion and other high-temperature steps in the entire CMOS process.   As shown in the figure, the final lightly-doped ($< 5 \times 10^{17}$ cm^{-3}) thickness is only 2 μm.   This effect must be taken into account when designing well depth in building CMOS on an epi film.

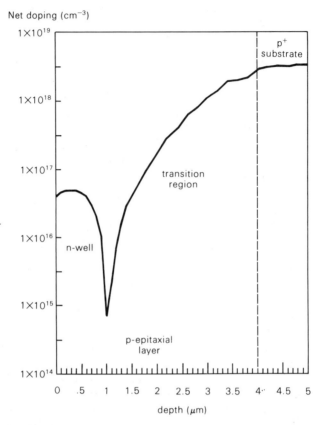

Figure 3.9 A p-on-p^+ epi doping profile and diffused n-well profile measured after the entire CMOS processing. The as-grown epi thickness is 4 μm and the final lightly-doped epi thickness is about 2 μm (Ref. 16, © 1987 IEEE).

3.2.5 Lithography and Etching

In IC fabrication, lithography and etching are two processes by which IC patterns are transferred from a set of masks to silicon wafers. The lithography process involves transferring an IC pattern from a mask to a photosensitive polymer film, called photoresist. The etching process replicates the photoresist pattern onto the underlying layer of a silicon surface. The photoresist layer can also be used as a masking layer to restrict ion implantation in certain areas of a silicon substrate. In either case, photoresist is stripped afterward and is not a part of the finished silicon wafer.

Lithography. The lithography process can be divided into two parts: mask making and photoresist operation. The mask making process starts after an integrated circuit is designed. Geometries of various IC components (such as gates and contacts) and interconnects are drawn and displayed using CAD (Computer-Aided Design) tools.   A composite drawing (often referred to as layout) of an IC is broken into levels for IC processing.   These are called masking levels with each level corresponding to a group of components made by a specific IC processing: polysilicon gates for one level, contact holes for another, and so on.   The multi-level IC layout is then digitized to drive a computer-controlled pattern generator which is often an electron beam machine.   It can transfer all the geometries of a particular level onto a glass plate covered with hard-surface material such as chromium.   After masking plates for all the levels are made, the whole set of plates is transferred to a photoresist room for pattern transfer.

The photoresist room contains an exposure tool to transfer patterns from the masking plate onto a silicon wafer.   This exposure tool also aligns the mask and the wafer so that the mask pattern, when transferred to the wafer, can be aligned with the previous pattern already defined on the wafer. The exposure tool is often characterized by pattern resolution and alignment accuracy.   Resolution is the minimum line or spacing that can be resolved. A state-of-the-art exposure tool using UV light has 1 μm resolution, i.e., 1 μm lines spaced by 1 μm in a reasonably thick (about 1 μm) photoresist. Its alignment accuracy is approximately ± 0.5 μm at the 3σ point where σ is the standard deviation of the statistical distribution of alignment errors.

As shown in Fig. 3.10, wafers are first spin-coated by photoresist and pre-baked before loading onto the exposure tool.   The wafers are then aligned and exposed by UV light, one by one.   The light goes through a mask before it reaches the wafer.   After exposure, the wafer is soaked in a solution (called developer), which dissolves either exposed area or unexposed parts.   This development process is very similar to photographic development.

There are two types of photoresists: positive resist and negative resist. If a positive resist is used, the exposed portion of the resist is dissolved in the developer; whereas for a negative resist, it is the unexposed portion that is dissolved.   At present, positive resists are most commonly used for high resolution lithography.   After development, wafers are post-baked for better adhesion during subsequent processes such as etching or implantation.   An inspection procedure is normally included to ensure acceptable resolution and alignment accuracy.

Etching.   Resist patterns printed by lithography must be transferred onto the layers that comprise the integrated circuit.   A common process is to selectively etch away the layer material not covered by the resist.   The etching process can be used to etch contact holes in an oxide layer or etch most polysilicon material away to leave narrow islands for MOSFET gates. Conventional etching processes use a chemical solution to dissolve unmasked portions of the layer in a way similar to the photoresist development process. The chemical etchant should not attack the photoresist mask, nor should it

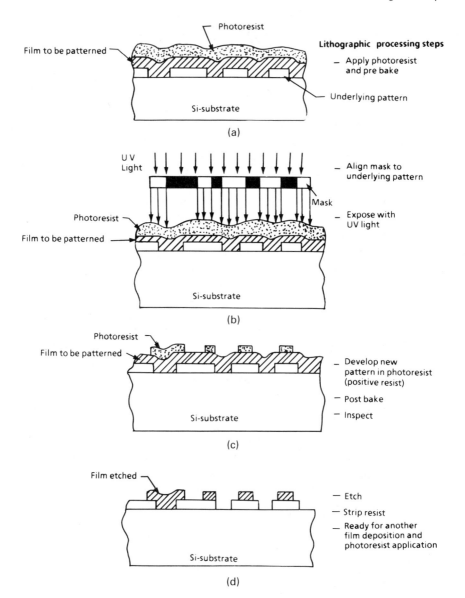

Figure 3.10 Processing steps in lithography and etching.

etch the material under the layer to be etched. The nature of this type of wet etching is isotropic so that etching occurs laterally as well as vertically. As a result, etched layers undercut the resist mask as shown in Fig. 3.11. Notice that the sidewall of the etched polysilicon gate is not vertical and its dimension is not well defined. This undercut problem becomes more severe

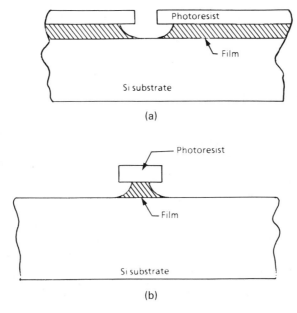

Figure 3.11   Undercut profiles due to etching: (a) contact hole etching profile; (b) polysilicon gate etching profile.

if overetching occurs because etching is not stopped according to time.   To alleviate this problem, dry etching has been introduced to avoid the use of wet chemical etchant.

Different types of dry etching are known as sputtering etching, plasma etching, reactive ion etching and reactive ion beam etching with all of them using plasmas in the form of low-pressure gaseous discharge.   They are now commonly used in IC processing because of their ability to precisely transfer high resolution patterns from a photoresist image to an underlying layer.

For a high quality etching, anisotropy and selectivity are the two most important factors.   An anisotropic etching can often be achieved by physically removing material using energetic ions.   The resultant profile of the etched material has vertical sidewalls with minimum undercut.   Sputtering etching, using 500-eV argon ions, is in this category.   Ar ions are accelerated to a cathode on which a sample is clamped.   These ions bombard the sample at normal incidence resulting in etching occurring in a vertical direction only. In this case, a high degree of anisotropy is inherent, but etching selectivity for different materials is poor.   Unless etching can be stopped in time precisely, etching continues to the underlying material after the layer needed to be etched has already been removed.   To avoid this problem, a chemical reaction is introduced during dry etching.   Plasma etching is an example by which molecular gas discharges serve as reactive species which react with the material of the layer to be etched. Once this layer is removed, reaction with the underlying material is minimal due to the high selectivity inherent in the nature of chemical etching. Reactive ion etching (RIE) also uses a molecular gas discharge, but in an apparatus similar to a sputtering etcher.

Both chemical and physical reactions occur during etching. With an appropriate gas flow rate, relatively low pressure, and adequate machine configuration, a high degree of anisotropy and selectivity can be obtained during RIE. Other important issues are etching throughput, sample load effects, uniformity of the etch rate across a wafer, and radiation damage. The CMOS gate oxide and the silicon/oxide interface are susceptible to radiation damage caused by the energetic particles present in plasma. Fortunately, most damage can be removed by a high-temperature anneal. However, introducing this anneal after aluminum deposition and patterning must be limited to temperatures below 500°C to avoid aluminum–silicon interaction. The energy of the particles must be kept low enough so that etching-induced damage can be annealed at a lower temperature (400–500°C).

Models have been developed for lithography and etching. A computer program for Simulation And Modelling of Profiles for Lithography and Etching (SAMPLE) was designed[17,18] to simulate line-edge profiles at key stages of IC processing. The simulation includes resist profiles produced by optical lithography, and polysilicon or other CVD-deposited film profiles etched by plasma etching or RIE. This user-oriented program is a useful tool for developing lithographic and etching processes.

3.3 MODELS FOR PROCESS INTEGRATION

3.3.1 One-dimensional Model

In VLSI, impurity profiles in the channel region are complicated due to shallow and multiple implants. A one-dimensional process integration program, SUPREM[9,10] has been developed and is widely used in the IC industry. This program primarily calculates impurity distributions but also computes oxide thickness and other process parameters. It accounts for any high temperature-induced diffusion. Impurity redistribution at oxidation, annealing, and diffusion are taken into account. The primary output of the program is the one-dimensional impurity distribution at various positions, e.g., the channel region and source/drain regions of a device. The program was designed to simulate process steps individually or sequentially just as they occur in an actual IC fabrication process. The individual process models previously described are employed and integrated in this program.

The third version, SUPREM III,[19] is capable of modeling multilayer structures up to five layers. The additional layers above the bulk and/or oxide are nitride and polysilicon. The thickness, dopant profile, and resistivity in the case of polysilicon are calculated.   Upgraded diffusion, oxidation, ion implantation and other individual process models are incorporated.   For example, a Si vacancy model is included to simulate the enhanced oxidation rate above n^+ and p^+ material.   SUPREM III also produces better prediction for the thin-oxide region.   Moreover, polysilicon modeling, including deposition, oxidation, doping segregation and the resulting resistivity, is included in the revised program.   To expedite the understanding between process sequence and device characteristics, SUPREM III calculates net charge, conductivity, and sheet resistance associated with electrical layers.   It also uses a one-dimensional Poisson's equation to obtain device parameters such as threshold voltage.

A typical example, shown in Fig. 3.12(a), is the input file for simulating a p-channel FET formed in the n-well of a CMOS IC.   The comment statements are for the users only.   After specifying the substrate type and resistivity, the user sets up grids and grid size.   Numerical solutions in general become more accurate with denser grids, but require more computing time.   SUPREM allows variable grids, meaning denser grids can be applied to the region where significant variation is expected.   SUPREM contains built-in oxidation models, wet or dry, with various pressures.   However, these models can be modified by users as needed.   The output can be tabulated or plotted normally as impurity versus vertical location from Si surface to substrate (Fig. 3.12b). n-Type, p-type or net impurity profiles can be calculated and displayed.   Notice the net impurity concentration changes from p- to n-type, indicating a p-type layer in this buried-channel FET.   The profile can be stored in a file for plotting or for use as the input file for device models such as GEMINI and MINIMOS which will be subsequently discussed.   Because the required computation can be done rapidly, the user can experiment with different processes by changing one or two process step(s) to examine the effects on the entire process.

3.3.2 Two-dimensional Model

For small sized MOS transistors, two-dimensional effects in the impurity distribution play an important role in device modelling.   Several two-dimensional (2-D) IC process models have been proposed and 2-D impurity profiles have been simulated.[20-22]   Among them, a 2-D process model, SUPRA,[22] has become commercially available and has been widely adapted.   SUPRA is a process simulator similar to SUPREM in many ways, but incorporates a 2-D model.   The 2-D diffusion equations discussed previously form the basis of this model.   It can handle local oxidation, nonplanar surfaces, implantation through a mask, and enhanced diffusion.   A combination of analytical and numerical methods is used in this model.   Both ion implantation and diffusion with low concentrations are modelled analytically as described by the 2-D equations derived in the last section.   At high concentration levels, impurity

```
TITLE                SHALLOW CONVENTIONAL NWELL CMOS
INITIALIZE           < 100 > SILICON BORON = 8.4E14
+                    THICKNESS = 4 DX = 0.005 .XDX = 0.0 SPACES = 120

COMMENT              INITIAL DX
DEPOSIT              OXIDE THICKNESS = 0.032 SPACES = 10
DEPOSIT              NITRIDE THICKNESS = 0.1 SPACES = 50

COMMENT              NWELL IMPLANT
IMPLANT              PHOS DOSE = 8E12 ENERGY = 170 PEARSON
COMMENT              NWLL DRIVE 8 HOURS
DIFFUSION            TEMP = 1050   TIME = 20 DRYO2
DIFFUSION            TEMP = 1050   TIME = 495 NITROGEN

COMMENT              FIELD OXIDATION 6000A
DIFFUSION            TEMP = 900  TIME = 383 WETO2
DIFFUSION            TEMP = 900  TIME = 55 NITROGEN
ETCH                 NITRIDE
ETCH                 OXIDE

COMMENT              GOX 1 200A
DIFFUSION            TEMP = 800 TIME = 51 WETO2
DIFFUSION            TEMP = 800 TIME = 40 NITROGEN
COMMENT              BF2 EHN IMP
IMPLANT              BORON DOSE = 4E12 ENERGY = 15 PEARSON
PRINT                LAYER
ETCH                 OXIDE

COMMENT              GOX 2 210A
DIFFUSION            TEMP = 800  TIME = 87 WETO2
DIFFUSION            TEMP = 800  TIME = 40 NITROGEN
DEPOSIT              POLYSILICON PRESSURE = 0.5 PHOS = 1E 18
+                    THICKNESS = 0.3 TEMPERATURE = 600
COMMENT              PCO DIFFUSION
DIFFUSION            TEMP = 900 TIME = 65 DRYO2
ETCH                 OXIDE
COMMENT              REOX
DIFFUSION            TEMP = 800  TIME = 12 DRYO2
DIFFUSION            TEMP = 800  TIME = 40 NITROGEN
COMMENT              ANNEALING
DIFFUSION            TEMP = 950  TIME = 15 NITROGEN
SAVEFILE             FILENAME = NWELL 68 STRUCTURE
SAVEFILE             FILE NAME = NWELL68.S2D SUPREM2
STOP
END

$TY PLOT.68
COMMENT              PLOT
INITIALIZE           STRUCTURE = NWELL68
PLOT                 FILE = NWELL68 HP2623A RIGHT = 1.0  LEFT = 0
+                    BOTTOM = 1E14 TOP = 1E18
+                    CHEMICAL NET
END
```

(a)

Figure 3.12   The SUPREM (a) input file; and (b) the output plot of an example on the *p*-channel profile in an *n*-well CMOS process.

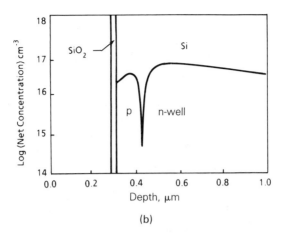

(b)

Figure 3.12 *Continued.*

diffusivity is no longer constant and the corresponding 2-D diffusion equation is solved numerically.   Concentration-enhanced diffusion is modelled through interaction with vacancies and clustering effects.[23]   Figure 3.13 shows an example in which impurity profiles under the field oxide as well as the ones in the channel region are simulated.   Notice that the relatively low-concentration boron profiles under the semi-recessed field oxide are calculated analytically whereas the arsenic profile associated with the S/D n^+ region is calculated numerically due to its high concentration.

The simulation of a 2-D impurity concentration in the region between n^+ to n-well in a CMOS structure[24] provides another example.   Figure 3.14(a) and (b) shows the SUPRA input file and the simulated 2-D equi-concentration

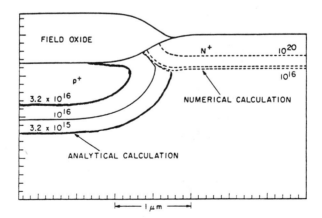

Figure 3.13 A SUPRA simulated 2-D profile of a cross-section of the test structure (Ref. 22).

```
TITLE        CMOS 4 N+ TO N-WELL

STRUCTURE    BORON=8E14 CONCENTRATION <100> WIDTH=7 DEPTH=4 HEIGHT=1.5
X.GRID       H1=0.1 H2=0.1 WIDTH=7 N.SPACES=70
Y.GRID       H1=.1 H2=.02 DEPTH=1.5 N.SPACES=15
Y.GRID       H1=0.02 H2=.1 DEPTH=4 N.SPACES=50
END

PRINT        SET SUMMARY

ANALYTIC

COMMENT      PAD OXIDE AND NITRIDE
DEPOSIT      OXIDE THICKNESS=0.020
DEPOSIT      NITRIDE THICKNESS=0.1

COMMENT      DEFINE SOME ACTIVE AREA
ETCH         NITRIDE X1=2 X4=7

COMMENT      N-WELL IMPLANTS
DEPOSIT      PHOTORES THICKNESS=5
ETCH         PHOTORES X1=3.2 X4=7
IMPLANT      PHOSPHORUS DOSE=4E12 ENERGY=170
IMPLANT      PHOSPHORUS DOSE=4E12 ENERGY=40
ETCH         PHOTORES X1=0 X4=7

COMMENT      WELL DRIVE-IN
OXIDIZE      TIME=480 TEMPERATURE=1050 INERT

COMMENT      BLANKET BORON FIELD IMPLANT
DEPOSIT      PHOTORES THICKNESS=5
ETCH         PHOTORES X1=2 X4=7
IMPLANT      BORON DOSE=2.5E12 ENERGY=140
ETCH         PHOTORES X1=0 X4=7

COMMENT      FIELD OXIDATION
OXIDIZE      TIME=5 TEMPERATURE=900 INERT
OXIDIZE      TIME=400 TEMPERATURE=900 WET
OXIDIZE      TIME=30 TEMPERATURE=900 INERT
ETCH         NITRIDE X1=0 X4=7

COMMENT NUMERICAL

REGRID       LOCAL SET

COMMENT      N+ AND CHANNEL IMPLANTS
IMPLANT      BORON DOSE=7E11 ENERGY=70
IMPLANT      ARSENIC DOSE=6E15 ENERGY=160

COMMENT      FINAL ANNEAL
OXIDIZE      TIME=30 TEMPERATURE=950 INERT

DEPOSIT      OXIDE THICKNESS=0.40
ETCH         OXIDE X1=0 X4=0.5
ETCH         OXIDE X1=6.5 X4=7.0

NUMERICAL
SAVE         FILE=SAVEFILE.SAV STRUCTURE
SAVE         FILE=GEMINI.GEM GEMINI
```

(a)

Figure 3.14   (a) The SUPRA input file; and (b) the resulting 2-D output plot of a test structure examining the n^+ to n-well isolation problem (Ref. 24, © 1987 IEEE).

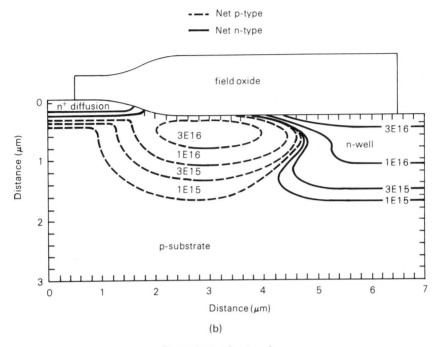

SIMULATED DOPING CONTOURS IN n^+ TO N-WELL PARASITIC MOSFET
NOMINAL SEPARATION = 2.4 MICRONS

Figure 3.14   *Continued.*

contours, respectively.   In this example, boron ions are implanted under the
entire field oxide to form a channel stop for *n*MOSFETs.   The *p*-type field
region inside the well is compensated for by the phosphorus implants used
during prior well formation.   Also, notice the simulated bird's beak at the
edge of the LOCOS field oxide.

3.4 TWO-DIMENSIONAL DEVICE MODELS

The impurity profiles, obtained from the 1-D or 2-D process model, do not
provide device characteristics.   Although 1-D MOS device physics are known
as described in Chapter 2, 2-D effects must be modelled accurately in small-
geometry devices.   Again, numerical simulation is a powerful tool in gaining
insights to a device in two dimensions.

3.4.1 Electrostatic Models

Various electrostatic models based on Poisson's equations have been
developed.[25-27]   Among them, the most popular has been the 2-D Poisson's

solver for nonplanar devices, known as GEMINI.[27] It provides electrostatic solutions assuming potential is not affected by current. Thus, potential (Φ) can be solved electrostatically using Poisson's equation alone:

$$\partial^2\Phi/\partial x^2 + \partial^2\Phi/\partial y^2 = -\rho(x,y)/\varepsilon_{si} \qquad (3.16)$$

where ρ is charge density and ε_{si} is the dielectric constant of silicon. This approximation is valid for linear and subthreshold regions in which only small amounts of current exist. It can also be applied to simulate punchthrough effects where current just starts to flow. After the electrostatic potential is obtained, currents are then calculated based on dependence to potential.

The current in the linear or subthreshold region is calculated as

$$I_d = g_d V_d \qquad (3.17)$$

where g_d is given as

$$
\begin{aligned}
g_d &= (Wd_{inv}/L)[\langle\rho\rangle]^{-1} = (Wd_{inv})[L\langle\rho\rangle]^{-1} \\[2mm]
&= (Wd_{inv})\left[\int_o^L \langle\sigma(x)\rangle^{-1}dx\right]^{-1} \\[2mm]
&= W\left[\int_o^L \left(\int_0^{d_{inv}} \sigma(x,y)dy\right)^{-1} dx\right]^{-1} \\[2mm]
&= qW\left[\int_o^L \left(\int_0^{d_{inv}} \mu_n(x,y)n(x,y)dy\right)^{-1} dx\right]^{-1}
\end{aligned}
\qquad (3.18)
$$

where d_{inv} is the inversion layer thickness, ρ and σ are the resistivity and conductivity, and $\langle\ \rangle$ means average value. $n(x,y)$ is the minority carrier concentration, and its value is based on the channel potential.

The punchthrough current calculation in GEMINI is much more involved. The model is based on the concept of drain-induced barrier lowering.[27] The current path can be at the surface or in the bulk depending on device structure and gate voltage. An extreme point P can be found along the current path as shown in Fig. 3.15. The calculated potential distribution around this point suggests that the potential at P is minimum along the r axis and maximum along the θ axis. Because the longitudinal electric field vanishes at P, the current is due entirely to minority carrier diffusion, and can be expressed as

$$J = -q\mu_n Wn(r,\theta)d\Phi_{Fn}(r)/dr \qquad (3.19)$$

where $\Phi_{Fn}(r)$ is the quasi-Fermi potential for the minority carriers, assuming electrons are here. The electron concentration at (r,θ) is

$$n(r,\theta) = (n_i^2/N_A)\exp(q\Phi^*/kT)\exp\{q[h(r,\theta) - \Phi_{Fn}(r)]/kT\} \qquad (3.20)$$

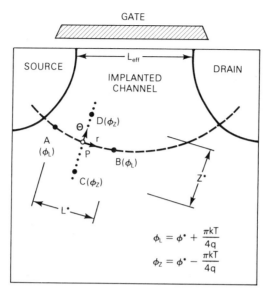

Figure 3.15 Cross-section of a MOSFET with a bulk current path showing the r and θ axes used to calculate the current. The points defining the approximate effective length L^* and effective width Z^* are shown (Ref. 27, © 1980 IEEE).

where Φ^* is the potential at P and $h(r,\theta)$ is the potential variation from Φ^*, i.e. $h(r,\theta) = \Phi(r,\theta) - \Phi^*$. Substituting Eq. 3.20 into Eq. 3.19, and separating r- and θ-dependent terms, the final I_d expression is

$$I_d = -qD_nW(Z^*/L^*)(n_i^2/N_A)\exp[q(\Phi^* - V_s)/kT]$$
$$\times \{1 - \exp -[q(V_d - V_s)/kT]\} \qquad (3.21)$$

where L^* and Z^* are the effective length and width of the electron path near the point P, which controls current. Although the detailed derivations of L^* and Z^* are described elsewhere,[27] their approximation is shown in Fig. 3.15. L^* is the distance along the r axis between points A and B where $\Phi = \Phi^* + \pi kT/4q$ and Z^* is the distance along the θ axis between C and D where $\Phi = \Phi^* - \pi kT/4q$. In the case of the current path at the surface, $Z^* = kT/(q|E_s|)$, where E_s is the vertical electric field at P, L^* remains unchanged. The approximation in GEMINI is justified by comparing its results favorably with CADDET,[29] which solves Poisson's and current continuity equations simultaneously. The calculated results also compare well with experimental data as shown in Fig. 3.16(a) and (b). Those calculations are for $W/L = 10$ μm/0.89 μm, $t_{ox} = 500$ Å, $x_j = 0.45$ μm, and $N_A = 2.8 \times 10^{15}$ cm^{-3}. The primary advantage of applying this Poisson's solver rather than current continuity equations is the significant reduction of computation time, hence extensive simulations can be done for device design and process

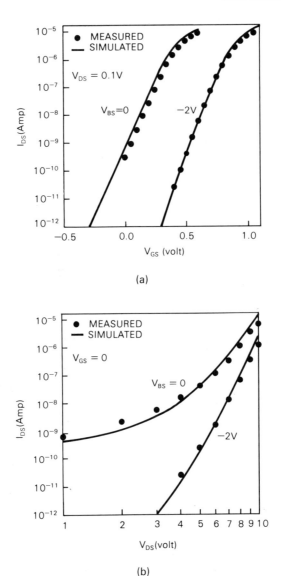

Figure 3.16 Measured and GEMINI simulated subthreshold currents for a short-channel MOSFET: (a) current vs. V_{GS} at V_{DS} = 0.1 V; and (b) current vs. V_{DS} at V_{GS} = 0 V. V_{BS} = V_{sub}, the substrate bias voltage (Ref. 27, © 1980 IEEE).

refinements. Numerically, GEMINI uses a 5-point approximation in solving finite difference equations in potential distribution. The program requires only a limited memory for execution and leaves the entire grid array matrix on a storage unit such as magnetic disc or tape. This implementation allows the simulation to be done with a minicomputer.

The GEMINI program takes impurity profiles from SUPREM, SUPRA or other sources and inputs them together with a device structure definition. In a device structure definition, oxide thickness, device geometry, gate electrode type and source/drain contacts are all specified.

Similar to SUPRA, the program sets up variable grids in two dimensions. Although the program is basically designed for simulating MOSFETs along the direction of channel length, it can also be applied for channel width simulation, parasitic field-oxide FETs, and other nonplanar structures such as a tapered oxide DRAM cell. The immediate output is the potential distribution in two dimensions, which is generally plotted out as 2-D equi-potential contours. Current can then be computed using the calculated potentials.

Using the example described in Sec. 3.3.2, a 2-D potential distribution can be calculated with the gate, source and drain electrodes defined and biases applied. Fig. 3.17 shows the input file and the corresponding 2-D equi-potential contours. The depletion edge is also shown. The IN.GEM file used for DATA.INP in the GEMINI input file is from the file saved from running SUPRA (see last statement in Fig. 3.14a).

3.4.2 Current Continuity Models

As described above, Poisson's solvers in GEMINI can only predict device operation at low current level. MOSFETs in saturation that have larger currents cannot be simulated by these models. Two-dimensional current continuity models coupled with Poisson's equation have been proposed. Restrictions in program application and computation time have limited their usage in device design. Recently, a user-oriented model, MINIMOS, has been widely used for the 2-D simulation of planar MOSFET's.[30]

MINIMOS solves current continuity equations and Poisson's equation iteratively, hence predicting current-voltage relations in the saturation region as well as in the linear and subthreshold regions. A finite difference method is applied for the following basic transport equations:

$$\nabla \cdot (\nabla \Phi) = -q(p - n + N_D - N_A) \quad \text{(Poisson's Eq.)} \tag{3.22}$$

$$\nabla \cdot \mathbf{J_n} - q \partial n / \partial t = qR \qquad \text{(Continuity Eq. for Electrons)} \tag{3.23}$$

$$\nabla \cdot \mathbf{J_p} + q \partial p / \partial t = -qR \qquad \text{(Continuity Eq. for Holes)} \tag{3.24}$$

$$\mathbf{J_n} = -q(\mu_n n \nabla \Phi - D_n \nabla n) \qquad \text{(Electron Current Eq.)} \tag{3.25}$$

$$\mathbf{J_p} = -q(\mu_p p \nabla \Phi - D_p \nabla p) \qquad \text{(Hole Current Eq.)} \tag{3.26}$$

The basic assumptions are: homogeneity of the dielectric constants, full ionization of all impurities, no bandgap narrowing, constant temperature over

```
COMMENT     N+ TO N-WELL MOSFET

STRUCTURE   TEMPERATURE=298 DATA.INP=IN.GEM

ELECTRODE   SOURCE ALUMINUM WIDTH=0.5
ELECTRODE   GATE ALUMINUM WIDTH=5 LEFTEDGE=1
ELECTRODE   DRAIN ALUMINUM WIDTH=0.5

END         STRUCTURE DEFINITION

SOLUTION    DATA.OUT=SAVEFILE.SAV

BIAS        SUBSTRATE POTENTIAL=0
BIAS        SOURCE POTENTIAL=5
BIAS        GATE POTENTIAL=5
BIAS        DRAIN POTENTIAL=5

END         SOLUTION DEF

COMMENTSTEP.VOL  MAXIMUM DELTA.VOL=2.0 STEPS=15 SUMM.OUT=STEP.STP
```

(a)

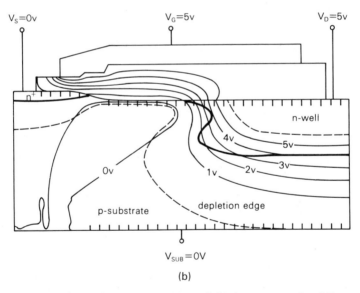

SIMULATED POTENTIAL CONTOURS IN n⁺ TO N-WELL PARASITIC MOSFET
NOMINAL SEPARATION=2.4 MICRONS

(b)

Figure 3.17 (a) The GEMINI input file; and (b) the corresponding 2-D output potential plot of an example examining the n^+ to n-well isolation problem (Ref. 24, © 1987 IEEE).

the device, only minority carriers contributing to current ($\mathbf{J_p} = 0$ for n-channel, and $\mathbf{J_n} = 0$ for p-channel) and no recombination and generation, i.e., $R = 0$. Substrate current can be modelled using an ionization integral.[31]

The features of the model include the calculation of impurity profiles and the modelling of channel mobility. Impurity profiles are calculated analytically within the model using process parameters such as implant conditions and temperature cycles. Profiles can also be input from SUPREM for vertical direction and can be fitted in the lateral dimension as well. The mobility model takes into account electrical field, impurity concentrations, temperature and the distance to the silicon/oxide interface. MINIMOS calculations can predict results for potential profiles, carrier distributions, and current densities, all in two dimensions.

PISCES[32] is a 2-D, two-carrier device model that simulates device characteristics in steady-state or transient conditions. It solves coupled Poisson's equation and current continuity equations. Thus, on-state devices such as a MOSFET under saturation can be modelled successfully. The program can handle arbitrary physical structures and general doping profiles obtained from SUPREM or analytic functions.

The basic transport equations used in PISCES are Equations (3.22) to (3.26). However, Eq. (3.25) and Eq. (3.26) co-exist because the 2-carrier model takes both electron and hole current into account. Also, physical models are more comprehensive and accurate. Additionally, Fermi–Dirac statistics, incomplete impurity ionization, bandgap narrowing, and Schottky barrier lowering associated with gate material are included. The mobility is modelled by[30]

$$\mu(E) = \mu_o \left[\frac{1}{1 + (\mu_o E/v_{\mathrm{sat}})\beta'} \right]^{1/\beta'} \tag{3.27}$$

where v_{sat} is the saturation velocity, μ_o is the low-field mobility, and β' is a constant between 1 and 2. The recombination rate is modelled by both Shockley-Read-Hall[33] and Auger band-to-band recombination.[34] Moreover, PISCES supports surface recombination using surface velocity. Dirichlet boundary conditions are applied on ohmic, Schottky and insulator contacts. On an ohmic contact, the minority and majority carrier quasi-Fermi potentials are equal and are set to the terminal bias. Schottky or insulator contacts are defined by the work function of the electrode. Surface recombination velocity and barrier lowering are taken into account for Schottky contacts. Electron and hole concentrations are forced to zero for insulator contacts. Neumann boundary conditions are applied on the outer edges rather than the contacts. There, the normal component of the electric field is either zero or equal to surface charge density, if it is present.

Because the primary limitation of using a 2-D model for design is computer time, non-uniform triangular grids are used. Furthermore, the initial coarse grids defined based on a device's structure can be refined before or during the solution process according to the doping profile or any other physical variable. As an example, Fig. 3.18 shows the coarse and fine grids defined for an n-well CMOS device with trench isolation.[35] PISCES is coded in finite differences on irregular grids using generalized box discretization. However, in the box method, each partial differential equation is integrated

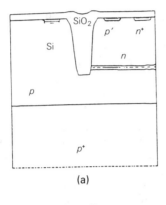

(a)

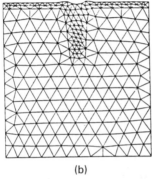

(b)

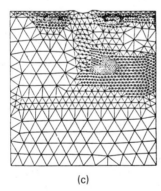

(c)

Figure 3.18 PISCES simulation on a trench isolated CMOS process. (a) Cross-section; (b) initial grid; and (c) refined working grid (Ref. 35, © 1985 IEEE).

over a small polygon enclosing each node. PISCES uses a finite element approach to sum equations on an element-by-element basis.[35]

The numerical method used in PISCES is based on the Gummel and Newton methods. The Gummel method[36] decouples Poisson's equation and current continuity equations and solves them sequentially and iteratively. This method is preferred for zero/reverse bias and low-current conditions such as the subthreshold region in a MOSFET. However, at high-current injection, especially high drift current in which the equations are strongly coupled, convergence is slow. The Newton method[37] is more suitable in this case

```
Title           HMOS FET – drain characteristics

$  .....        Load mesh
mesh            infile = hmesh

$  .....        Symbolic factorization
*symb           gummel carriers = 1 electrons
method          iccg damped

$  .....        Materials/contacts
mater           num = 2  g.sur f = 0.75
contac          num = 1  n.poly

$  .....        Models
models          conmob temp = 300 fldmob print

$****************************************************
$  .....        Load initial solution from hout0
load            infile = hout0

$  .....        Increment gate voltage (VGS = 1.2)
solve           v1 = 1  vstep = 1  nsteps = I electrode = 1

$  .....        Setup IV log file
log             cutfile = IV.drain

$  .....        Solve for VGS = 3 volts. save in hout3
solve           vl = 3  outfile = hout3

$  .....        Switch methods
*symb           newton carriers = 1  electrons
method          autonr

$  .....        Step VDS from 0.2 to 3.0 volts (VGS = 3 volts) and
$  .....        output solutions to houtdx
solve           v4 = 0.2  vstep = 0.2  nsteps = I4 electrode = 4 outfile = houtda

$  .....        Plot ID vs VDS
plot.1d  x.axis = v4 y.axis = I4

end
```

(a)

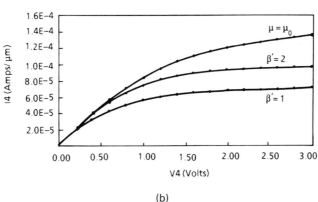

(b)

Figure 3.19   (a) PISCES input file; and (b) the resulting I-V curves at $V_G = 3$ V for three different mobility models (Ref. 32).

because all the coupling variables are taken into account through a ninefold increase in matrix size. Inverting this large matrix however takes a large amount of computer memory and time. This method is chosen for a MOS-FET after turn-on, particularly in the saturation region; it is also required for transient and ac analyses.[32] Figure 3.19 shows an example of a PISCES simulation for a MOSFET biased into saturation region. Notice the switch to the Newton method when the drain voltage steps from 0.2 to 3 volts. The I-V curves correspond to three different mobility models all for $V_G = 3$ V.

3.5 DEVICE PARAMETER EXTRACTION FOR CIRCUIT SIMULATION

All models discussed so far are based on physical principles. There are, however, parameters such as mobility which have a strong empirical dependence, and their values cannot be accurately derived from first-principle physics. Therefore, it is more adequate to extract these values from transistor measurements.

3.5.1 Parameter Extraction Algorithm

In a typical transistor test, parameters such as threshold, transconductance, punchthrough voltage, etc., are measured sequentially and separately. The danger is that the values of these parameters are assumed constant and accurate for obtaining subsequent parameters. For example, it is inappropriate to use the mobility measured in a linear region to obtain the saturation current in a short-channel MOSFET. A computer-aided parameter extraction program, SUXES (Stanford University eXtractor of modEl parameterS), which extracts a complete set of parameters simultaneously by a single operation, has been developed.[38] The Laverberg-Marquart[39] algorithm is used to extract the complete parameter set by performing a least-square fit of the model to measured data. Rather than extracting a single parameter at a time, the method extracts a vector ($\mathbf{p}$) which represents a set of parameters through an optimization algorithm. The objective function to be optimized is $f(\mathbf{p})$ and it is expressed as[40]

$$\| f(\mathbf{p}) \|^2 = \mathbf{\Sigma}_k f_k(\mathbf{p})^2 = \mathbf{\Sigma}_k \left[\frac{I_k(\mathbf{p}) - I_k^*}{\max(I_k^*, I_{\min})} \right]^2 \qquad (3.28)$$

where I_k and I_k^* are the calculated and the measured drain current at the kth data points, respectively. $I_{\min}$ is the user-defined current used to normalize the error between the calculated and measured values. A block diagram of the optimization program is shown in Fig. 3.20.[41] This general-

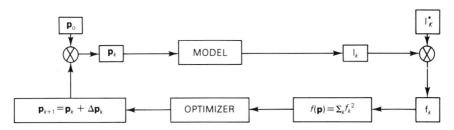

Figure 3.20   Block diagram for the SUXES parameter extraction algorithm (Ref. 41, © 1983 IEEE).

purpose optimization method is independent of the MOSFET model used. Thus, new models can be incorporated conveniently.

The results of a typical example using this extraction method are shown in Fig. 3.21 in which measured data are compared with the calculated I-V curve. Notice that the optimized result (solid curves) shows a good fit with the measured data (crosses) although the initial calculations (dashed curves) substantially depart from the data.

3.5.2 MOSFET Models for Extraction

The SPICE circuit simulator contains several MOSFET models. Table 3.2 shows the simplest, LEVEL 1 model,[42,43] which includes various capacitances (to be subsequently discussed) as well as DC parameters. Redundant parameters are included for providing flexibility to users. For example, the transconductance parameter (KP) depends on channel mobility, which is a function of the bias voltages because they determine the electric field. This

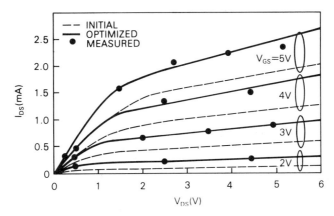

Figure 3.21   Drain characteristics of a 5-μm nMOSFET, showing good fit produced by optimized parameter extraction (Ref. 40, © 1982 IEEE).

TABLE 3.2 SPICE Level 1 MOS Transistor Model Parameters (Ref. 42).

Symbol	Name	Parameter	Units	Default	Example
	LEVEL	Model index		1	1.0
V_{T0}	VTO	Zero-bias threshold voltage	V	0.0	1.0
k'	KP	Transconductance parameter	A/V²	2.0E-5	3.1E-5
γ	GAMMA	Bulk threshold parameter	$V^{1/2}$	0.0	0.37
$2|\phi_F|$	PHI	Surface potential	V	0.6	0.65
λ	LAMBDA	Channel-length modulation	1/V	0.0	0.02
r_d	RD	Drain ohmic resistance	Ω	0.0	1.0
r_s	RS	Source ohmic resistance	Ω	0.0	1.0
C_{bd}	CBD	Zero-bias B-D junction capacitance	F	0.0	2.0E-14
C_{bs}	CBS	Zero-bias B-S junction capacitance	F	0.0	2.0E-14
I_s	IS	Bulk junction saturation current	A	1.0E-14	1.0E-15
ϕ_0	PB	Bulk junction potential	V	0.8	0.87
	CGSO	Gate-source overlap capacitance per meter channel width	F/m	0.0	4.0E-11
	CGDO	Gate-drain overlap capacitance per meter channel width	F/m	0.0	4.0E-11
	CGBO	Gate-bulk overlap capacitance per meter channel width	F/m	0.0	2.0E-10
	RSH	Drain and source diffusion sheet resistance	Ω/square	0.0	10.0

Symbol	Name	Parameter	Units		
C_{j0}	CJ	Zero-bias bulk junction bottom capacitance per square meter of junction area	F/m²	0.0	2.0E-4
m	MJ	Bulk junction bottom grading coefficient		0.5	0.5
	CJSW	Zero-bias bulk junction sidewall capacitance per meter of junction perimeter	F/m	0.0	1.0E-9
m	MJSW	Bulk junction sidewall grading coefficient		0.33	
	JS	Bulk junction saturation current per square meter of junction area	A/m²		1.0E-8
t_{ox}	TOX	Oxide thickness	m	1.0E-7	1.0E-7
N_A or N_D	NSUB	Substrate doping	1/cm³	0.0	4.0E-15
Q_{ss}/q	NSS	Surface state density	1/cm²	0.0	1.0E-10
	NFS	Fast surface state density	1/cm²	0.0	1.0E-10
	TPG	Type of gate material: +1 opposite to substrate −1 same as substrate 0 Al gate		1.0	
X_j	XJ	Metallurgical junction depth	m	0.0	1.0E-6
L_D	LD	Lateral diffusion	m	0.0	0.8E-6
μ	UO	Surface mobility	cm²/V · s	600	700

field-dependent mobility is not explicitly expressed in LEVEL 1. In LEVEL 2, the mobility is modelled based on the derivation described in Chapter 2. The mobility equation (Eq. 2.21) can be modified to

$$U = U_o\{\varepsilon_{si}U_{crit}/[C_{ox}(V_g - V_t - \text{UTRA} \times V_d)]\}^{\text{UEXP}} \qquad (3.29)$$

where the V_d term is added to account for the inversion charge loss due to pinchoff at high V_d. UTRA can be chosen between 0 and 0.5.[44]

More sophisticated MOS models have been developed to account for short channel effects in a MOSFET. The CSIM (Compact Short-channel IgFET Model),[45] later modified to BSIM (Berkeley Short-channel IgFET Model),[46] is an example of parameter extraction. The model is based on the device physics of small MOSFETs. Short-channel effects such as field-dependent mobility, drain-induced barrier lowering, charge sharing, and non-uniform doping are included. The original CSIM (CSIM1) is modelled using eight parameters:

V_{FB}, the flat-band voltage

Φ_s, the surface potential

γ, the body effect coefficient

K, the source/drain depletion charge-sharing coefficient

η, the drain-induced barrier lowering coefficient

β_o, the intrinsic conductance

U_o, the mobility degradation coefficient

U_1, the velocity saturation coefficient

The first five parameters model the threshold voltage and the remaining three parameters model the drain current. The associated equations can be found in Appendix 3.1.

The last four parameters (η, β_o, U_o, and U_1) depend on bias voltages (V_d and V_{sub}). They are expanded to 13 parameters in the CSIM2 model as described in Appendix 3.2.

The CSIM2 model is further improved and renamed as BSIM. In addition to the 17 parameters, three new parameters are added to model the subthreshold region in which $V_g < V_t$. As shown in Eq. (2.36), the subthreshold swing S_t is expressed as

$$S_t = (2.3kT/q)n = (2.3kT/q)(1 + C_d/C_{ox}) \qquad (3.30)$$

where n is the subthreshold slope coefficient. As discussed in Chapter 2 and shown in Fig. 2.15 and Fig. 2.22, the subthreshold swing is a function of substrate and drain biases. In BSIM, this bias dependence is modelled by a linear relation using three parameters[48]

$$n = n_o + n_s V_{sub} + n_d V_d \qquad (3.31)$$

•••BSIM EXTRACTION STATUS•••

PROCESS=CMOS2   VDD=5.00 VOLTS
LOT=182   TEMP=25.00 DEG C
WAFER=17   TOX=285.00 ANGSTROMS
DATE=4/18/85   XPOS=127 YPOS=127
OPERATOR=BW   DEVICE=NCHANNEL
OUTPUT FILE=W17_N2   WIDTH=20.00 MICRONS
PROBER FILE=SINGLE DEVICE OPERATION   LENGTH=2.00 MICRONS

MINUTES TO DIE COMPLETION=0.1   MINUTES TO WAFER COMPLETION=0.1
DEVICE EXTRACTION LOCATION XXXXXXXXXXXXXXXXXXXXXXXXXXXXXXXXXXXXXXX FINISHED
PRESENT DEVICE BSIM PARAMETERS
VFB=−0.799   X2U0=−0.000857
PHIF2=0.698   X2U1=−0.010352
K1=1.316   X3U1=0.029285
K2=0.289   X2BETA0=0.000000
ETA=−0.001   X2ETA=0.005606
BETA0=0.000637   X3ETA=0.001180
U0=0.084   BETA0SAT=0.001178
U1=0.300   X2BETA0SAT=−0.000010
N0=   X3BETA0SAT=0.000063
XZNB=   X3ND=
Are you interested in subthreshold region measurements and extraction?(Y/N) >

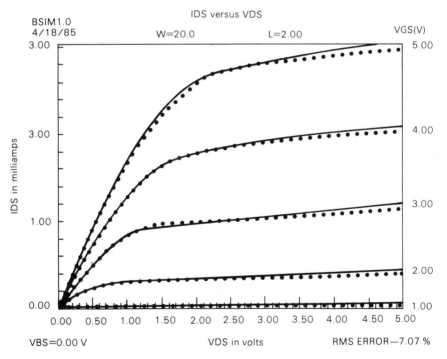

Figure 3.22   Output of the BSIM parameters and the resulting I-V characteristics for a 2-μm nMOSFET: measured, ●: calculated by BSIM.

The proposed model results in a smooth drain current from the subthreshold to the linear region. Channel shortening (ΔL) and channel narrowing (ΔW) are modelled through the intrinsic conductance (β_o) as

$$\beta_o \text{ (at } V_d = 0) = \mu_o C_{ox}(W_m - \Delta W)/(L_m - \Delta L) \qquad (3.32)$$

where W_m and L_m are mask dimensions and μ_o is the low-field mobility. By measuring a set of MOSFETs with various W_ms and L_ms on a test chip, one can extract ΔW from β_o vs. W_m, and ΔL from β_o^{-1} vs. L_m. All β_o values correspond to μ_o values based on the relation shown in Eq. 3.32.

The SPICE LEVEL 4 model is BSIM. The BSIM parameters are shown in the following table:

SPICE Name	BSIM Name	Parameter Description
VFB	V_{FB}	flat-band voltage
PHI	Φ_s	surface potential
K1	γ	body effect coefficient
K2	K	source/drain depletion charge-sharing coefficient
ET	η_z	zero-bias drain-induced barrier lowering coefficient
MUZ	μ_o	zero-bias mobility
DL	ΔL	channel shortening
DW	ΔW	channel narrowing
U0	U_{oz}	zero substrate bias mobility degradation coefficient
U1	U_{1z}	zero substrate bias velocity saturation coefficient
X2U0	U_{ob}	sensitivity of mobility degradation to substrate bias
X2U	U_{1b}	sensitivity of velocity saturation to substrate bias
X3U1	U_{1d}	sensitivity of velocity saturation to drain bias at $V_d = V_{dd}$
X2E	η_b	sensitivity of drain-induced barrier lowering to substrate bias
X3E	η_d	sensitivity of drain-induced barrier lowering to drain bias at $V_d = V_{dd}$
X2MZ	μ_{zb}	sensitivity of mobility to substrate bias at $V_d = 0$
MUS	μ_s	mobility value at zero substrate bias at $V_d = V_{dd}$
X2MS	μ_{sb}	sensitivity of mobility to substrate bias at $V_d = V_{dd}$
X3MS	μ_{sd}	sensitivity of mobility to drain bias at $V_d = V_{dd}$

To extract the above parameters, it is obvious that temperature at the parameters that are extracted (TEMP), gate oxide thickness (TOX) and measurement bias range (VDD) must be specified. MOSFET length (L) and width (W) should also be input. Figs. 3.22 and 3.23 show the printout of the BSIM extraction program for 2 μm n- and p-channel MOSFETs, respectively. The input data, the 17 extracted BSIM parameters, and the measured and calculated I-V curves are shown. BETA0, X2BETA0, BETA0SAT, X2BETA0SAT, and X3BETA0SAT correspond to MUZ, X2MZ, MUS, X2MS, and X3MS, respectively. Subthreshold parameters are not extracted in this example.

BSIM EXTRACTION STATUS

PROCESS=CMOS2	VDD=5.00 VOLTS
LOT=182	TEMP=25.00 DEG C
WAFER=17	TOX=285.00 ANGSTROMS
DATE=4/18/85	XPOS=127 YPOS=127
OPERATOR=BW	DEVICE=PCHANNEL
OUTPUT FILE=W17_P1	WIDTH=20.00 MICRONS
PROBER FILE=SINGLE DEVICE OPERATION	LENGTH=2.00 MICRONS

MINUTES TO DIE COMPLETION=0.1   MINUTES TO WAFER COMPLETION=0.1
DEVICE EXTRACTION LOCATION XXXXXXXXXXXXXXXXXXXXXXXXXXXXXXXXXXXXX FINISHED
PRESENT DEVICE BSIM PARAMETERS

VFB=−0.189	X2U0=0.005256
PHIF2=0.698	X2U1=−0.002813
K1=0.476	X3U1=0.001011
K2=0.014	X2BETA0=0.000009
ETA=−0.007	X2ETA=0.001248
BETA0=0.000293	X3ETA=0.000094
U0=0.179	BETA0SAT=0.000411
U1=0.085	X2BETA0SAT=0.000009
N0=	X3BETA0SAT=0.000011
X2NB=	X3ND=

Are you interested in subthreshold region measurements and extraction?(Y/N) >

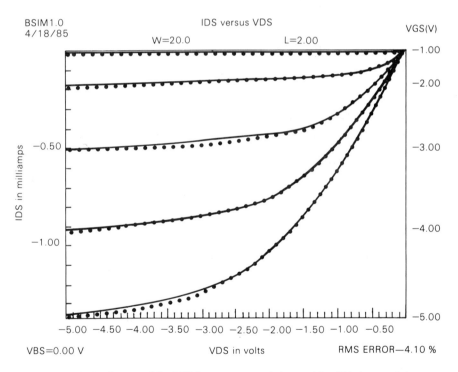

Figure 3.23 Output of the BSIM parameters and the resulting I-V characteristics for a 2-µm pMOSFET: measured, ●: calculated by BSIM.

3.5.3 MOSFET Capacitances for SPICE

For circuit simulation, accurate capacitances are necessary for predicting transient response and speed.   Several types of capacitance are important for MOS devices and circuits.   In SPICE 2G, the total junction capacitance (C_{jtot}) in a MOSFET is modelled as[44]

$$C_{jtot} = AS \times C_j[1 + V_R/V_{bi}]^{-MJ} + PS \times C_{JSW}[1 + V_R/V_{bi}]^{-MJSW} \qquad (3.33)$$

where V_R is the reverse-biased voltage at the source or drain with respect to the substrate, V_{bi} is the built-in voltage of the p-n junction, and AS and PS are the junction bottom area and junction sidewall perimeter, respectively. The other terms are SPICE input parameters and are expressed as CJ or C_j, the zero-bias junction capacitance per m^2 of junction area; CJSW or C_{jsw}, the zero-bias junction capacitance per m of junction sidewall perimeter; MJ, the grading coefficient of the junction bottom; and, MJSW, the grading coefficient of the junction sidewall.   MJ and MJSW are 1/2 if a step junction is assumed and are 1/3 if a linearly-graded junction is modelled.[3]

In addition to the junction capacitances, the SPICE BSIM model has four capacitance parameters with three corresponding to overlap capacitances, and one associated with partitioning the channel charge in calculating intrinsic capacitances.   For short-channel MOSFETs, the capacitance due to physical overlap (l_{ov}) between gate to source or gate to drain becomes a significant portion of the total gate oxide capacitance, and hence cannot be ignored (Fig. 3.24a).   Additionally, the fringing field capacitance from the gate to source or drain is not negligible due to the finite thickness associated with the polysilicon gate.   Fig. 3.24(b) shows that the proportion of the fringing field capacitance (C_{fr}) to the total oxide capacitance (C_{ox}) increases as the gate length is decreased.[49]   The analytic solution shown in the figure is derived using conformal mapping and Laplace's equation.   Therefore, even if a VLSI process results in zero lateral diffusion (shown in Fig. 3.24b), the total overlap capacitance ($C_{ov} + C_{fr}$) is non-zero due to electrical overlapping originated from the fringing field.   These overlapping capacitances can also be measured from test devices.[50]   The three overlap capacitance parameters used in the SPICE/BSIM[51] are:

CGDO or C_{gdo}, gate-drain overlap capacitance per meter channel width,

CGSO or C_{gso}, gate-source overlap capacitance per meter channel width,

CGBO or C_{gbo}, gate-bulk overlap capacitance per meter channel length.

The gate-bulk overlap capacitance (C_{gbo}) is the field-oxide capacitance resulting from the gate extending beyond the channel to ensure a proper overlap of the field oxide.   The total gate-bulk overlap capacitance is the C_{gbo} multiplied by the channel length (L).   The gate-drain and the gate-source overlap

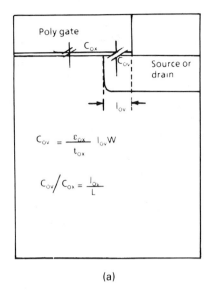

$$C_{ov} = \frac{\varepsilon_{ox}}{t_{ox}} \, l_{ov} W$$

$$C_{ov}/C_{ox} = \frac{l_{ov}}{L}$$

(a)

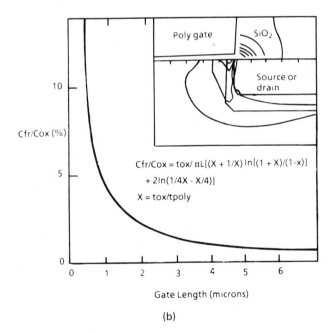

$$Cfr/Cox = tox/\,\pi L[(X + 1/X)\,\ln[(1 + X)/(1-x)]$$
$$+ \, 2\ln(1/4X - X/4)]$$
$$X = tox/tpoly$$

Gate Length (microns)

(b)

Figure 3.24   The effect of finite gate thickness on the gate-to-drain and gate-to-source capacitance; (a) C_{ov} is the overlapping capacitance and (b) C_{fr} is the fringe-field capacitance (Ref. 49).

capacitances described above are in units of per channel width. C_{gdo} and C_{gso} are the constant portions of the total C_{gd} and C_{gs}, which also contain capacitances due to voltage-dependent charges associated with the gate and channel. The voltage-dependent charge when differentiated by one of the terminal voltages is often referred to as the intrinsic capacitance.

Intrinsic capacitances are modelled by Ward and Dutton[52] based on Mayer's model considering capacitance as the variation of charge with a terminal voltage.[53] Mathematically, the MOSFET capacitances are defined as

$$C_{ij} = \partial Q_i / \partial V_j \qquad (3.34)$$

where i and j denote any two out of the four terminals in a MOSFET. For example, $C_{gd} = \partial Q_g / V_d$ while all other terminal voltages are held constant. The MOSFET has a total of 12 capacitances, but only 6 out of the 12 are independent terms due to charge conservation:

$$Q_{channel} = Q_d + Q_s = -(Q_g + Q_b). \qquad (3.35)$$

These capacitance terms are used to relate the current flowing into the terminal and the charge contained in the associated region. The transient current-voltage relation is solved numerically using Newton-Raphson iteration[52] and the algorithm is employed in SPICE 2.G.[44] To a first order approximation, the intrinsic capacitances from gate to other terminals add to a constant:

$$C_g = C_{gs} + C_{gd} + C_{gb} \approx C_{ox} = A\varepsilon_{ox}/t_{ox} \qquad (3.36)$$

where C_{ox} is the capacitance of the MOS gate oxide. C_{gs}, C_{gd} and C_{gb} are dependent on the distribution of the channel charge at source, drain and bulk.

In SPICE 2.G, a parameter named XQC indicates the portion of the channel charge attributed to drain. XQC is specified between 0–0.5. In SPICE 3, BSIM is used to model intrinsic capacitances again through charge conservation.[46] The model integrates distributed charges over the active area. Two methods are used to partition channel charge into drain and source components. The source/drain partitioning of channel charge ranges from 50/50 in the linear region to 100/0 (or 60/40 in the second method) in the saturation region. The associated BSIM parameter in SPICE 3 is X_{part}, representing the following partitioning factor:

XPART or X_{part}, gate-oxide capacitance charge model flag.

$X_{part} = 0$ selects a 60/40 source/drain charge partition in saturation, while $X_{part} = 1$ sets a 100/0 partition in source/drain. $X_{part} = 1$ is normally

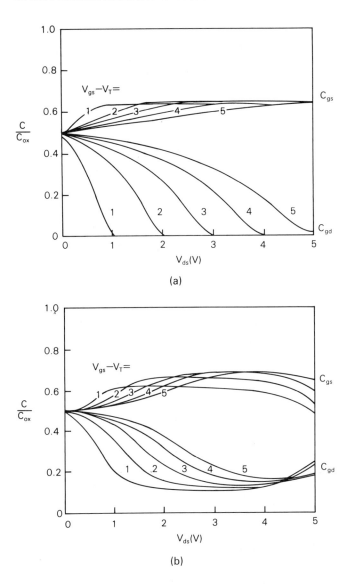

Figure 3.25   Intrinsic capacitances in a MOSFET: (a) long-channel; and (b) short-channel (Ref. 46).

recommended.   Figure 3.25(a) shows the intrinsic capacitances of a long-channel device versus V_{ds} for several V_{gs} values.   In the saturation region where $V_{ds} \geq V_{gs} - V_t$, $C_{gd} = 0$ and $C_{gs} = (2/3)C_{ox}$ because all the channel charges are at source and the drain has no charge due to pinchoff.   In the linear region where $V_{ds} \ll V_{gs} - V_t$, C_{gs} and C_{gd} approach the value of

$(1/2)C_{ox}$. In the cut-off region where $V_{gs} - V_t < 0$, $C_{gd} = C_{gs} = 0$ and $C_g = C_{gb} = C_{ox}$ because there is no channel formed.

Recently, a charge-based analytical model was developed to include short-channel effects such as mobility degradation, channel-length modulation, bias-dependent fringing field effect and source/drain series resistance.[54] The gate charge Q_g is composed of three components,

$$Q_g = Q_{g1} + Q_{g2} + Q_{g3} \qquad (3.37)$$

where Q_{g1} is the charge in the source region where gradual-channel approximation applies, Q_{g2} is the charge in the drain region where carrier velocity saturates, and Q_{g3} is the charge corresponding to the channel-side bias-dependent fringing field capacitance. Analytic expressions are derived for all three components and capacitances C_{gd} and C_{gs} are calculated. Comparisons of simulated results to measured values are satisfactory for 1 μm devices. A typical result is shown in Fig. 3.25(b). Notice that C_{gd} does not decrease to zero in the saturation region because the short channel is not quite pinched off. This enhancement will be included in revised versions of BSIM models.[55]

3.5.4 Parasitic Capacitances, Resistances and Inductances

So far, capacitances associated with a MOSFET device have been discussed. Other parasitic capacitances and resistances are also important in circuit simulation. The parasitic capacitances and resistances associated with a one-micron MOSFET in a VLSI environment is shown in Fig. 3.26,[56] which is drawn roughly to scale. Notice the large capacitive coupling between interconnects-to-device, and among interconnects themselves. All of these capacitances in conjunction with layout areas must be accounted for when simulating circuit performance. The inter-wiring capacitance at tight metal pitches increases as the metal wire spacing is decreased and eventually becomes a dominant component in total wiring capacitance.

Inductive noise due to current pulse is also important, especially for CMOS circuits, because a spike of current occurs during switching. Figure 3.27 shows the typical line inductance on a chip and the noise caused by an inverter.[56] Notice that the line inductance increases as the line width is decreased. The parasitic inductance in I/O's that normally switch large currents should be included in circuit simulation.

Also worth mentioning is the effect of chip topology on parasitic capacitances and resistances. An increase in interlayer capacitance at steps is significant and is a function of layer profiles. As shown in Fig. 3.28, the capacitance and resistance of a metal line passing over the poly gate increase as the underlying topology becomes more severe,[57] due primarily to the PSG (Phosphor-Silica-Glass) or Al thinning at the polysilicon sidewall.

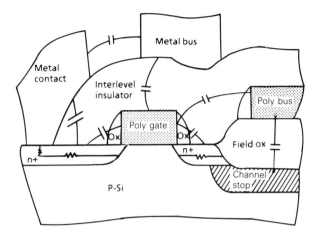

Figure 3.26   Scale drawing of the parasitic environment in 1 μm MOS technology (Ref. 56).

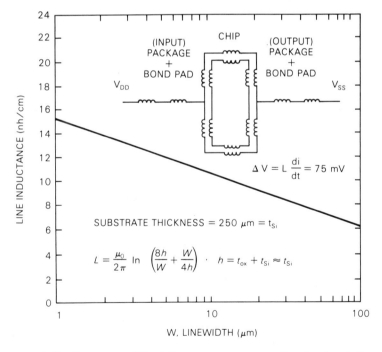

Figure 3.27   The impact of line inductance on power supply noise. Insert shows noise due to a typical switching of a CMOS driver (Ref. 56).

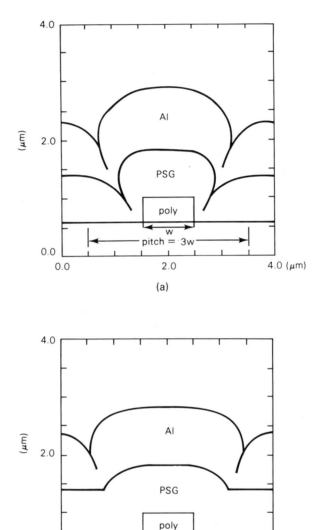

(a)

(b)

Figure 3.28 Simulated profiles of PSG and Al lines over polysilicon gates: (a) without reflow, (b) with reflow, and (c) the resulting capacitance and resistance values. The polysilicon is 0.4 μm high covered by 0.8-μm PSG and then 1.0-μm Al (Ref. 57, © 1983 IEEE).

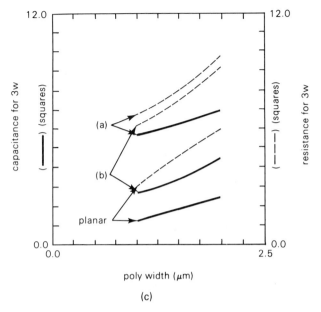

Figure 3.28 *Continued.*

APPENDIX 3.1   CSIM1 MODEL PARAMETERS

The CSIM1 model consists of eight parameters that describe the characteristics of a MOSFET.   The first five parameters model the threshold voltage as

$$V_t = V_{FB} + \Phi_s + \gamma\sqrt{\Phi_s - V_{sub}} - K(\Phi_s - V_{sub}) - \eta V_{ds} \qquad (A3.1)$$

V_{FB}, Φ_s and γ have been described in Chapter 2. K and η are the parameters that model the threshold lowering effect. η is roughly inversely proportional to the channel length (L).[47]   The remaining three parameters (β_o, U_o and U_1) model the drain current as follows:

In the cut-off region ($V_g < V_t$)

$$I_d = 0 \qquad (A3.2)$$

In the linear region ($V_g > V_t$ and $V_d < V_{dsat}$), carrier mobility can be simplified to

$$U = \frac{U_o}{(1 + U_1 V_d)[1 + U_o(V_g - V_t)]} \qquad (A3.3)$$

and the commonly observed 3/2 power dependence term in the I-V relation is approximated numerically.[47]   The current can then be written as

$$I_d = \frac{\beta_1}{(1 + U_1 V_d)} [(V_g - V_t) - a V_d/2] V_d \qquad (A3.4)$$

where
$$\beta_1 = \beta_o[1 + U_o(V_g - V_t)]^{-1} \tag{A3.4a}$$

and
$$\beta_o = \mu_o(W/L)C_{ox} \tag{A3.4b}$$

Also,
$$a = 1 + \frac{\gamma}{2\sqrt{\Phi_s - V_{sub}}}\left[1 - \frac{1}{1.744 + 0.8364(\Phi_s - V_{sub})}\right] \tag{A3.4c}$$

In the saturation region ($V_g > V_t$, and $V_d > V_{dsat}$):

$$I_d = \frac{\beta_1}{2aV}(V_g - V_t)^2 \tag{A3.5}$$

where

$$V = \frac{1 + v_c + \sqrt{1 + 2v_c}}{2}, \quad \text{and} \quad v_c = U_1(V_g - V_t)/a \tag{A3.6}$$

APPENDIX 3.2　BIAS-DEPENDENT MODEL PARAMETERS IN CSIM2

The last four parameters in the CSIM model (β_o, U_o, U_1, η) are functions of drain and substrate voltages and can be expressed by

$$U_o = U_{oz} + U_{ob}V_{sub} \tag{A3.7}$$

$$U_1 = U_{1z} + U_{1b}V_{sub} + U_{1d}(V_d - V_{dd}) \tag{A3.8}$$

$$\eta = \eta_z + \eta_b V_{sub} + \eta_d(V_d - V_{dd}) \tag{A3.9}$$

β_o is modelled by quadratic interpolation through three data points, at $V_d = 0$, $V_d = V_{dd}$, and the slope of β_o to V_d at $V_d = V_{dd}$. It can be expressed as

$$\beta_o = \beta_o(\text{at } V_d = 0)(V_d/V_{dd} - 1)^2 + \beta_o(\text{at } V_d = V_{dd}) \tag{A3.10}$$
$$\times (2 - V_d/V_{dd})V_d/V_{dd} + \beta_{sd}V_d(V_d/V_{dd} - 1)$$

where

$$\beta_o(\text{at } V_d = 0) = \beta_z + \beta_{zb}V_{sub} \tag{A3.11}$$

$$\beta_o(\text{at } V_d = V_{dd}) = \beta_s + \beta_{sb}V_{sub} \tag{A3.12}$$

$$\beta_o(\text{at } V_{sub} = 0) = \beta_s + \beta_{sd}(V_d - V_{dd}) \tag{A3.13}$$

V_{sub} in the above equations is the absolute value of substrate voltage, V_d is the drain voltage and V_{dd} is the supply voltage.   Other parameters are

U_{oz}, the U_o value at zero substrate bias,

U_{ob}, the sensitivity of U_o to substrate bias,

U_{1z}, the U_1 value at zero substrate bias,

U_{1b}, the sensitivity of U_1 to substrate bias,

U_{1d}, the sensitivity of U_1 to drain bias at $V_d = V_{dd}$

η_z, the η value at zero substrate bias and $V_d = V_{dd}$,

η_b, the sensitivity of η_o to substrate bias,

η_d, the sensitivity of η to drain bias at $V_d = V_{dd}$

β_z, the β_o value at zero substrate and drain biases,

β_{zb}, the sensitivity of β_o to substrate bias at $V_d = 0$,

β_s, the β_o value at zero substrate bias and $V_d = V_{dd}$,

β_{sb}, the sensitivity of β_o to substrate bias at $V_d = V_{dd}$,

β_{sd}, the sensitivity of β_o to drain bias at $V_d = V_{dd}$.

REFERENCES

1. J. F. Gibbons, W. W. Johnson and S. W. Mylroie, Projected range statistics in semiconductors, John Wiley & Sons, 1975.

2. Process and device simulation for MOS-VLSI circuits, ed. by P. Antognetti, D. A. Antoniadis, R. W. Dutton and W. G. Oldham, Martinus Nijhoff Publishers, 1983.

3. A. S. Grove, Physics and technology of semiconductor devices, John Wiley & Sons, 1967.

4. VLSI Technology, ed. by S. M. Sze, McGraw-Hill, 1983.

5. W. K. Hofker, "Implantation of boron in silicon," *Philips Research Reports, Suppl.*, No. 8, 1975.

6. S. Selberherr, Analysis and simulation of semiconductor devices, Springer-Verlag, Wien, New York, 1984.

7. Re-arranged from the tables in reference 1.

8. S. Furukawa, H. Matsumura, and H. Ishiwara, "Lateral distribution theory of implanted ions," in *Ion Implantation in Semiconductors*, ed. by S. Namba, p. 73, 1972.

9. K. Lehovec and A. Slobodskoy, "Diffusion of charged particles into a semiconductor under consideration of the built-in field," *Solid-State Electron.*, vol. 3, p. 45, 1961.

10. D. A. Antoniadis, S. E. Hansen, R. W. Dutton, and A. G. Gonzales, "SUPREM I—A program for IC process modeling and simulation," Stanford Electronics Lab., Stanford University, Stanford, CA, *Tech. Rep.* 5019-1, May 1977.

11. D. A. Antoniadis and R. W. Dutton, "Models for computer simulation of complete IC fabrication process," *IEEE Trans.* on *Electron Devices*, ED-26, p.490, 1979.

12. B. E. Deal and A. S. Grove, "General relationship for the thermal oxidation of silicon," *J. Appl. Phys.*, vol. 36, p. 3770, 1965.

13. R. B. Fair and J. C. C. Tsai, "Theory and direct measurement of boron segregation in SiO_2 during dry, near dry and wet O_2 oxidation," *J. Electrochem. Soc.*, vol. 125, p. 2050, 1978.

14. A. S. Grove, O. Leistiko, and C. T. Sah, "Redistribution of acceptor and donor impurities during thermal oxidation of silicon," *J. Appl. Phys.*, vol. 35, p. 2695, 1964.

15. A. S. Grove, A. Roder, and C. T. Sah, "Impurity distribution in epitaxial growth," *J. Appl. Phys.*, vol. 36, p. 802, 1965.

16. A. G. Lewis, R. A. Martin, T-Y Huang, J. Y. Chen, and M. Koyanagi, "Latchup performance of retrograde and conventional *n*-well CMOS technologies," *IEEE Trans. on Electron Devices*, vol. ED-34, p. 2156, 1987.

17. W. G. Oldham, S. N. Nandgaonkar, A. R. Neureuther, and M. O'Toole, "General Simulator for VLSI lithography and etching processes: part I—Application to projection lithography," *IEEE Trans. on Electron Devices*, vol. ED-26, p. 717, 1979.

18. W. F. Oldham, A. R. Neureuther, C. Sung, J. L. Reynolds, and S. N. Nandgaondar, "General Simulator for VLSI lithography and etching processes: part II—Application to deposition and etching," *IEEE Trans. on Electron Devices*, vol. ED-27, p.1455, 1980.

19. C. P. Ho, J. D. Plummer, S. E. Hansen, and R. W. Dutton, "VLSI process modelling—SUPREM III," *IEEE Trans. on Electron Devices*, vol. ED-30, p. 1438, 1983.

20. R. Tielert, "Two-dimensional numerical simulation of impurity redistribution in VLSI processes," *IEEE J. Solid-State Circuits*, vol. SC-15, p. 544, 1980.

21. H. Ryssel, K. Haberger, K. Hoffmann, G. Prinke, R. Dumcke, and A. Sachs, "Simulation of doping processes," *IEEE J. Solid-State Circuits*, vol. SC-15, p. 549, 1980.

22. D. Chin, M. Kump, and R. W. Dutton, "SUPRA—Stanford University PRocess Analysis program," Stanford Report, Stanford Electronics Laboratories, Stanford University, Stanford, July, 1981.

23. D. J. Chin, M. R. Kump, H. G. Lee, and R. W. Dutton, "Process design using coupled 2D process and device simulators," *IEDM Digest*, p. 223, 1980.

24. A. G. Lewis, J. Y. Chen, R. A. Martin, and T. Y. Huang, "Device isolation in high density LOCOS-isolated CMOS," *IEEE Trans. Electron Devices*, vol. ED-34, p. 1337, 1987.

25. D. J. Coe, J. E. Brockman, and K. H. Nicholas, "A comparison of simple and numerical two-dimensional models for the threshold voltage of short channel MOSFETs," *Solid-State Electronics*, vol. 20, p. 993, 1977.

26. D. B. Scott and S. G. Chamberlain, "A calibrated model for the subthreshold operation of a short channel MOSFET including surface states," *IEEE J. Solid State Circuits*, vol. SC-14, p. 633, 1979.

27. J. A. Greenfield and R. W. Dutton, "Nonplanar VLSI device analysis using the solution of Poisson's equation," *IEEE Trans. Electron Devices*, ED-27, p. 1520, 1980.

28. R. R. Troutman, "VLSI Limitations From Drain-Induced Barrier Lowering," *IEEE Trans. Electron Device*, ED-26, p. 461, 1979.

29. T. Toyable and S. Asai, "Analytical models of threshold voltage and breakdown voltage of short-channel MOSFETs derived from two-dimensional analysis," *IEEE Trans. Electron Devices*, vol. ED-26, p.453, 1979.

30. S. Selberherr, A. Schutz, and H. W. Potzl, "MINIMOS-A two-dimensional MOS transistor analyzer," *IEEE J. Solid-State Circuits*, vol. SC-15, p. 605, 1980.

31. H. Oka, K. Nishiuchi, T. Nakamura, and H. Ishikawa, "Two-dimensional analysis of buried channel MOSFETs" in *IEDM Digest*, 1979.

32. M. R. Pinto, C. S. Rafferty, and R. W. Dutton, "PISCES-II: Poisson and continuity equation solver," Stanford Electronics Lab. Tech. Rep., Sept. 1984.

33. R. N. Hall, "Electron-hole recombination in germanium," *Phys. Rev.*, vol. 87, p. 387, 1952.

34. W. Shockley and W. T. Read, "Statistics of the recombination of holes and electrons," *Phys. Rev.*, vol. 87, p. 835, 1952.

35. C. S. Rafferty, M. R. Pinto, and R. W. Dutton, "Iterative methods in semiconductor device simulation," *IEEE Trans. on Electron Devices*, vol. ED-32, p. 3018, 1985.

36. D. L. Sharfetter and H. K. Gummel, "Large-signal analysis of a silicon Read diode oscillator," *IEEE Trans. Electron Devices*, vol. ED-16, p. 64, 1969.

37. B. V. Ghokale, "Numerical solutions for a one-dimensional silicon *n-p-n* transistor," *IEEE Trans. Electron Devices*, vol. ED-17, p. 594, 1970.

38. K. Doganis, R. W. Dutton, and A. G. Gonzalez, "Optimization of IC process using SUPREM," Stanford Electronics Lab., Stanford Univ., Tech. Rep., June, 1981.

39. K. M. Brown and J. E. Dennis, Jr., "Derivative free analogues of Levenberg-Marquadt and Gauss algorithms for nonlinear least-squares approximation," *Numerical Math.*, vol. 18, p. 289, 1972.

40. D. E. Ward and K. Doganis, "Optimized extraction of MOS model parameters," *IEEE Trans. on Computer-Aided Design of Integrated Circuits and Systems*, vol. CAD-1, p. 163, 1982.

41. K. Doganis and D. L. Scharfetter, "General optimization and extraction of IC device model parameters," *IEEE Trans. Electron Devices*, vol. ED-30, p. 1219, 1983.

42. D. A. Hodges and H. G. Jackson, Analysis and Design of Digital Integrated Circuits, p. 56, McGraw Hill, 1983.

43. H. Shichman and D. A. Hodges, "Modeling and simulation of insulated-gate field-effect transistors," *IEEE J. of Solid-state Circuits*, vol. SC-3, p. 285, 1968.

44. A. Valadimirescou and S. Liu, "The simulation of MOS integrated circuits using SPICE 2," Univ. of Calif. Berkeley, CA, Memo. UCB/ERL M80/7, Feb. 1980.

45. B. J. Sheu, D. L. Sharfetter, and H. C. Poon, "Compact short-channel IGFET model (CSIM)," Univ. of Calif. Berkeley, CA, Memo. UCB/ERL M84/20, March 1984.

46. B. Sheu, D. L. Sharfetter, and P. K. Ko, "BSIM, an IC process-oriented MOSFET model and the associated characterization system," Proceeding of ISCAS, p. 433, 1985.

47. H. C. Poon, "Vth and beyond," Workshops on Device Meeting for VLSI, Burlingame, CA, March, 1979.

48. A. H.-C. Fung, "A subthreshold conduction model for BSIM," Univ. of Calif., Berkeley, Memo. UCB/ERL M85/22, March 1985.

49. P. Chatterjee, P. Yang, and H. Shichijo, "Modelling of small MOS devices and device limits," *IEE Proc.*, vol. 130, p. 105, 1983.

50. R. Shrivastava and K. Fitzpatrick, "A simple model for the overlap capacitance of a VLSI MOS device," *IEEE Trans. on Electron Devices*, vol. ED-29, p. 1870, 1982.

51. T. Quarles, A. R. Newton, D. O. Pederson, A. Sangiovanni- Vincentelli, "SPICE 3A5 User's Guide," EECS Dept., Univ. of Calif., Berkeley, Nov. 1985.

52. D. E. Ward and R. W. Dutton, "A charge-oriented model for MOS transistor capacitances," *J. of Solid State Circuits*, vol. SC-13, p. 703, 1978.

53. J. E. Mayer, "MOS models and circuit simulation," RCA Rev., vol. 32, p. 42, 1971.

54. B. J. Sheu and P. K. Ko, "An analytical model for intrinsic capacitances of short-channel MOSFETs," *IEDM Digest*, p. 300, 1984.

55. P. K. Ko, Private communication.

56. P. Chatterjee, "Device design issues for deep submicron VLSI," Proc. of Inter'l Symp. on VLSI Technology, Systems, and Applications, Taipei, Taiwan, p. 221, 1985.

57. K. Lee, Y. Sakai, and A. R. Neureuther, "Topography-dependent electrical parameter simulation for VLSI design," *IEEE Trans. Electron Devices*, vol. ED-30, p. 1469, 1983.

EXERCISES

1. When designing a process for the formation of p-channel source/drain in the n-well of a CMOS, if the n-well concentration is 5×10^{16} cm^{-3} and is covered by an oxide of 250 Å, the lowest boron energy you can operate from your implanter is 10 KeV and the total dose is 5×10^{14} cm^{-2}. Estimate what will be the as-implanted source/drain junction depth.

2. For the boron implant profile obtained in Exercise 1, if the wafers are annealed at 1000°C for 1/2 hour, taking into account boron diffusion in extrinsic Si, calculate the new junction depth.

3. If photoresist mask is used for an n-well implant that uses 8×10^{12} cm^{-2} phosphorus

ions at 170 KeV and the p-type substrate is 5×10^{15} cm^{-2}, how thick must the resist be to serve as a mask?

4. Based on the example in Fig. 3.14, run SUPRA with boron doses at 1×10^{12}, 5×10^{12} and 1×10^{13} cm^{-2} for the blanket boron field implant and examine the resulting 2-D impurity distributions between the n-well and n^+.

5. Using the three output files in Exercise 4, run GEMINI to observe equi-potential contours for three different boron doses.

6. Explain the differences between MINIMOS and PISCES in terms of equations used, basic assumptions and limitations.

7. For a MOSFET with $W = 10\ \mu$m, $L = 1\ \mu$m, $t_{ox} = 200$ Å, t_{poly} (poly-gate thickness) = 3000 Å, l_{ov} (source/drain to gate overlap) = 0.1 μm, and a poly gate overlap field oxide by 2 μm, calculate the total gate-to-drain (C_{gd}), gate-to-source (C_{gs}) and gate-to-bulk (C_{gb}) capacitances (including overlap capacitances) in the linear, saturation and cutoff regions.

8. If a 1 mm-long 1 μm-wide CMOS metal line switches 10 mA in 1 ns, calculate the voltage noise caused by the parasitic line inductance (use Fig. 3.27).

4

CMOS OPERATION

In this chapter, the CMOS operation is introduced through a description of simple CMOS circuit elements. These elements are important components in CMOS VLSI circuits. The objective is to provide some ideas about how circuit operation is affected by the technology. The emphasis is CMOS circuits; however, they will be compared to typical NMOS circuits. The chapter starts with an introduction of a CMOS inverter, a CMOS output buffer and a CMOS transmission gate because they are the basic components. Then, some simple components for logic and memory circuits are described.

4.1 INVERTERS

4.1.1 An NMOS Inverter

An inverter is the most basic element in building logic circuits. A typical NMOS technology normally uses an enhancement-mode FET together with a depletion-mode FET to form a so-called E-D NMOS inverter as shown in Fig. 4.1a. The lower device is the enhancement-mode FET used as a driver, whereas the depletion-mode FET serves as a load. Both devices are made in a p-type substrate and the substrate is normally grounded or sometimes

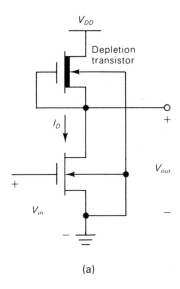

(a)

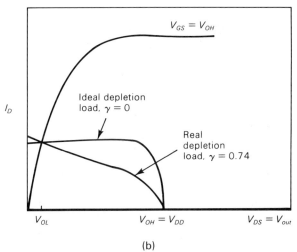

(b)

Figure 4.1   A depletion load NMOS inverter: (a) the schematic; and (b) the I-V characteristics (Ref. 2).

biased at a negative voltage by a substrate generator.   The source and gate of the depletion device are connected to give zero V_{gs}, thereby resulting in a nearly constant current source.   The load line for the output I-V of the E-D inverter is shown in Fig. 4.1b.   For the load device, its V_{ds} (hence, I_{ds}) increases as V_{out} is decreased.   An ideal load device should act as a constant current source.   In reality, lower current is normally observed because of the associated body effect. As the V_{out} (i.e. V_s of the load) increases, V_{TD} (threshold of the depletion-mode device) increases because the substrate-to-source is negatively biased. As a result, V_{TD} at $V_{sub} = 0$ is set lower, i.e., more negative, to provide the same logic low level (V_{OL}) as shown in the figure. Notice that the load line deviates from an ideal current source as the body effect coefficient (γ) changes from 0 to 0.74. Though the logic high level at output (V_{OH}) is always at V_{DD}, the V_{OL} value depends on the W/L values for both load and driver devices. The low and high noise margins, N_{ML} and N_{MH}, are defined as

$$N_{ML} = V_{IL} - V_{OL} \tag{4.1a}$$

$$N_{MH} = V_{OH} - V_{IH} \tag{4.1b}$$

where V_{IL} and V_{IH} are the input voltages at the unity gain ($dV_{out}/dV_{in} = -1$) points.

As shown in the transfer curve of Fig. 4.2a, a smaller V_{OL} can be obtained by decreasing the Z (load)/Z (driver) ratio, where $Z = W/L$. However, a small Z ratio corresponds to a lower I_D as shown in Fig. 4.2b, hence slower operation. A good choice is to make the inversion voltage (V_{inv}), defined as the input voltage such that $V_{out} = V_{in}$, half-way between the V_{DD} and ground. This situation is met if the enhancement threshold $V_{TE} = 0.2$ V_{DD}, $V_{TD} = -0.6 V_{DD}$, and the Z (load) to Z (driver) ratio is approximately 1:4.[1,2]

4.1.2 A CMOS Inverter

The CMOS inverter shown in Fig. 4.3a consists of a p-channel load device and an n-channel driver; both are enhancement mode devices. The source and substrate of the p-channel are connected to V_{DD}, whereas the source and substrate of the n-channel are grounded. This arrangement ensures that $V_{bs} = 0$ for both devices. Thus, no body effect exists in the CMOS inverter. The input of the inverter (V_{in}) is connected to both p- and n-channel gates, and the drain areas of the two devices are also tied together for output (V_{out}). If the input is at V_{DD}, the n-channel device is on and the p-channel is off. When the input is switched to zero, the n-channel device is off and the p-channel load device is on because V_{gs} in p-channel is now at $-V_{DD}$.

As shown in Fig 4.3b, the current in the load device varies with the input voltage. Notice that $V_{dsn} = V_{out}$, but $V_{dsp} = V_{out} - V_{DD}$, V_{dsn} and V_{dsp} are the drain-to-source voltages for the n- and p-channel devices, respectively. The load line changes when the input switches from zero to V_{DD}, resulting in so-called dynamic load logic. Whether the input is at zero or V_{DD}, only one device is on, the other is off. As a result, a CMOS inverter offers an excellent noise margin because $V_{OH} = V_{DD}$ and $V_{OL} = 0$.

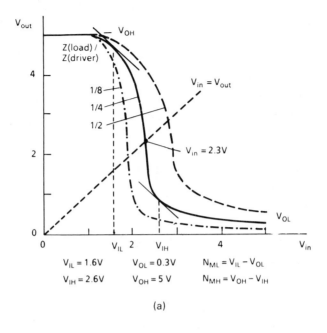

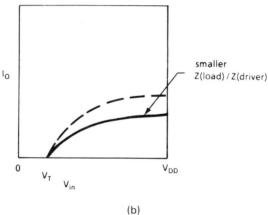

Figure 4.2 (a) DC transfer curves; and (b) output current vs. input voltage of depletion load NMOS inverters with various Z_{load}/Z_{driver} ratios, where $Z = W/L$ (Ref. 2).

The inverter transfer characteristic is shown in Fig. 4.4a. Notice that $V_{OL} = 0$ and is independent of the κ ratio, where $\kappa = (W/L)\mu\varepsilon_{ox}/t_{ox}$, and the κ ratio is

$$\frac{\kappa_n}{\kappa_p} = \frac{\mu_n W_n L_p}{\mu_p W_p L_n} \tag{4.2}$$

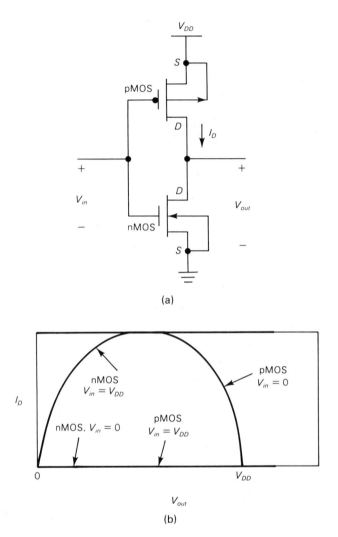

Figure 4.3   A CMOS inverter: (a) the schematic and (b) the I-V characteristics.

CMOS logic is therefore referred to as a ratioless logic.   For a κ ratio of unity, V_{inv} is equal to $V_{DD}/2$, giving the best result for both noise margins. Because $\mu_n \cong 2\mu_p$, to have $\kappa = 1$ and allow $L_p = L_n$, W_p is set to $2W_n$. Because one of the two devices is off at either logic state, essentially no steady-state current exists.   Only a small amount of steady-state current can exist due to the leakage current associated with the off-state device.   However, transient current exists when the V_{out} differs from 0 or V_{DD}.   In fact, when V_{in} switches between V_{tn} and $V_{DD} - |V_{tp}|$, neither transistor is off and

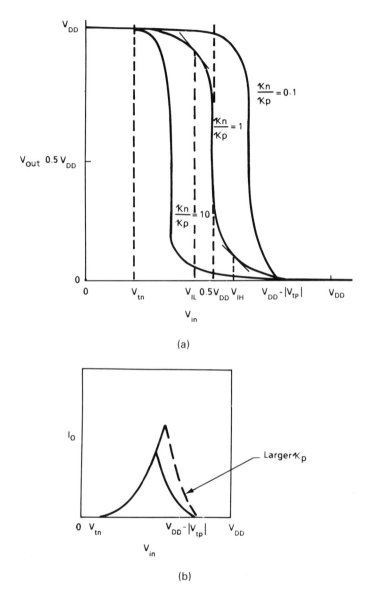

(a)

(b)

Figure 4.4   (a) DC transfer curves, and (b) output current vs. input voltage of CMOS inverters with various κ_n/κ_p ratios, where $\kappa = (W/L)\mu\varepsilon_{ox}/t_{ox}$ (Ref. 3).

a current spike flows at output as shown in Fig. 4.4b.   Unlike I_D in the NMOS inverter, the current here only exists during switching.   Detailed analyses and mathematical expressions are described elsewhere.[3]

4.1.3 Inverter Delay

When an inverter input steps from low to high, the amount of time for the output to change from V_{DD} to zero is called fall time (t_f).   Because it is the time needed to discharge the output node from V_{DD} to 0 with the n-channel current, t_f can be obtained by

$$\int_o^{t_f} I_{dn}(t)\ dt\ =\ C_L V_{DD} \tag{4.3}$$

where C_L is the load capacitance and I_{dn} is the n-channel drain current that increases from zero to the saturation current I_{dsat}.   Eq. 4.3 can then be approximated as

$$C_L V_{DD}\ =\ t_f I_{dsat}/2\ =\ t_f[\mu_n C_{ox}(W_n/L_n)(V_{DD}\ -\ V_{tn})^2/2]/2 \tag{4.4}$$

Solving t_f, the following is obtained

$$t_f\ =\ \frac{4C_L V_{DD} L_n}{\mu_n C_{ox} W_n (V_{DD}\ -\ V_{tn})^2} \tag{4.5a}$$

Similarly, the time required for output to charge from 0 to V_{DD} when the input steps from high to low can also be derived.   This time delay defined as rise time (t_r) or charging time because the output node has to be charged from 0 to V_{DD} by the p-channel device is

$$t_r\ =\ \frac{4C_L V_{DD} L_p}{\mu_p C_{ox} W_p (V_{DD}\ -\ |V_{tp}|)^2} \tag{4.5b}$$

As shown in Fig. 4.5, a more rigorous definition of t_f is the time from the input rising edge to 10% of the final output level. t_r is the time from the input falling edge to 90% of the final output level.   If the κ ratio is set to unity, i.e., $\mu_n W_n/L_n\ =\ \mu_p W_p/L_p$, the rise time equals the fall time and symmetric output waveform can be expected.   Both fall time and rise time are proportional to the load capacitance.

If the output of an inverter drives an identical inverter, then the load capacitance is the sum of the n-channel and p-channel gate capacitances $(C_{gn}$ and $C_{gp})$:

$$C_L\ =\ C_{gn}\ +\ C_{gp}\ =\ C_{ox}\ (W_n L_n\ +\ W_p L_p) \tag{4.6}$$

In IC fabrication, it is common to allow $L_n\ =\ L_p\ =\ L$, and because μ_n is approximately $2\mu_p$, $W_p\ =\ 2W_n$ can satisfy $\kappa\ =\ 1$.   Eq. 4.6 can then be written as

$$C_L=\ 3C_{ox} W_n L\ =\ (3/2)C_{ox} W_p L \tag{4.7}$$

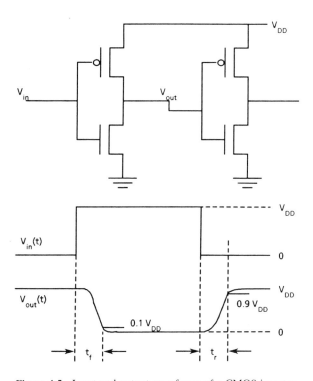

Figure 4.5   Input and output waveforms of a CMOS inverter.

Substituting Eq. 4.7 into Eq. 4.5 and assuming $V_{tn} = -V_{tp} = 0.2V_{DD}$, the following inverter delay time t_{in} can be obtained, defined as $(t_r + t_f)/2$:

$$t_{in} = t_f = t_r = 18.75 \, \tau_e = 9.375 \, \tau_h \qquad (4.8)$$

where

$$\tau_e = L^2/(\mu_n V_{DD}) \quad \text{and} \quad \tau_h = L^2/(\mu_p V_{DD}) \qquad (4.9)$$

τ_e and τ_h are the electron and hole transit times for the n- and p-channel devices, respectively. The condition $W_p = (\mu_n/\mu_p)W_n$ indicates identical rise time and fall time, but does not offer the minimum delay time. In fact, the inverter delay time is minimum if $W_p = \sqrt{\mu_n/\mu_p}W_n$. Because the delay time is proportional to the load capacitance, if an inverter drives f identical inverters as shown in Fig. 4.6, then the total load capacitance to be charged or discharged increases by a factor of f. As a result, the inverter delay time becomes ft_{in} for a fan-out factor of f.

Based on Eqs. 4.8 and 4.9, the inverter delay time, when driving another identical inverter, depends on a transistor's channel length, but not on its channel width. Both the load capacitance and transistor current scale up proportionally as channel width is increased. Experimental results however

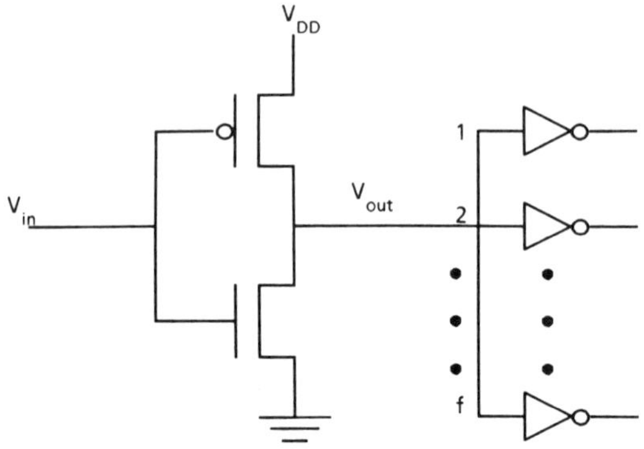

Figure 4.6   A CMOS inverter driving several inverters, fan-out $= f$.

do not quite reflect this relationship due to the parasitic effects that have not yet been considered in the approximation described above.

In an integrated circuit, capacitances at the inverter output node not only include the input capacitance of the gate(s) connected to it but also include other parasitic capacitances associated with the node. Figure 4.7 shows two significant parasitic capacitances: Miller capacitance (C_m) and stray capacitance (C_s). Every MOS transistor has the Miller capacitance, which is a capacitance between the drain edge of the gate and the drain node. This capacitance in an inverter string is charged to $-C_m V_{DD}$ when V_{o1} is low and V_{o2} is high, and is charged to $C_m V_{DD}$ when V_{o1} switches to high and V_{o2} goes low. Its effect on the circuit is therefore doubled because $[C_m V_{DD} - (-C_m V_{DD})]/V_{DD} = 2C_m$. The stray capacitance is due to the voltage swing of signal paths connected to a node. Signal paths are made of metal or polysilicon lines on thick insulator or field oxide as shown in Fig. 3.26. The associated line capacitances per unit area are at least an order of magnitude less than the gate capacitance per unit area; however, the area covered by

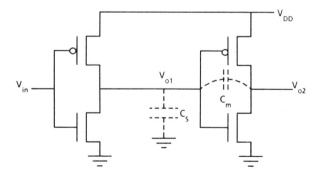

Figure 4.7   Node capacitances associated with a CMOS inverter pair.

these lines is often much larger than the gate area, causing a significant total line capacitance. In addition, as interconnect lines are packed closely together in VLSI, the line-to-line coupling capacitance adds another significant capacitance. The total stray capacitance of the signal lines may be large. Thus, the total node capacitance can be expressed as

$$C_{\text{total}} = C_{gn} + C_{gp} + 2C_m + C_s, \qquad (4.10\text{a})$$

and

$$C_m = W_n C_{gdo}(n\text{-channel}) + W_p C_{gdo}(p\text{-channel}) \qquad (4.10\text{b})$$

where C_{gdo}, gate-drain overlap capacitance per channel width, is defined in Sec. 3.5.3. If W_n and W_p double, I_{dn} and I_{dp} double and so do C_{gn}, C_{gp} and C_m. But because C_s remains basically unchanged, the increase of C_{total} is less than a factor of two. Because the transistor current doubles, the delay time will decrease and will decrease even more when C_s becomes larger due to more or longer interconnect lines.

4.2 A CMOS OUTPUT BUFFER

An output buffer is designed to deliver sufficient current to charge or discharge a large load capacitance, which consists of output pad, bonding wire, and pins of other chips. Figure 4.8 shows an NMOS and a CMOS buffer consisting of several stages with the last two being the pre-driver and driver. The NMOS driver is implemented by a typical push–pull configuration in which V_{out} follows input voltage of the pre-driver. In the case of a CMOS buffer, it is simply a chain of properly scaled inverters.

To drive a large load (C_L), a large inverter driver should be used to provide sufficient current. Because the internal inverter is small, it seems obvious that the internal inverter should drive a larger inverter, which would drive an even larger inverter until at some point the inverter is large enough to be the driver. Assuming that each inverter in the chain is larger than the preceding one by a factor of F, total delay is then nFt_{in} where n is the number of inverters and $F^n = C_L/C_g$, C_g is the internal capacitance.

To minimize the total delay, fewer inverters could be used but the delay from one inverter to the next would be longer. Or, a smaller F could also be used to reduce the individual inverter delay, but more inverters would become necessary, thus leading to an increase in total chain delay. It has been shown[1] that the total delay (nFt_{in}) is minimal if $F = e$, where e is the base of natural logarithms; and the number of stages is $\ln(C_L/C_g)$.

As previously described, scaling-up transistor widths for all inverters increases speed but the effect is not great. Figure 4.9 shows the transition of the output driver for different transistor widths.[5] It is clear that even doubling the width of the transistors does not substantially speed up the buffer

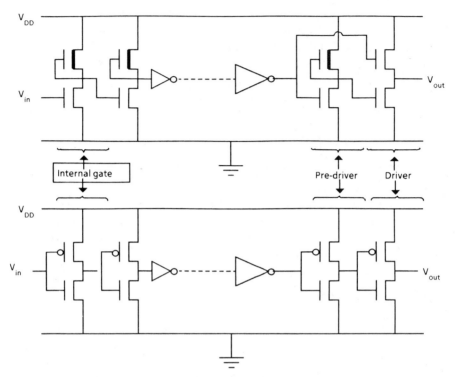

Figure 4.8 (a) An NMOS and (b) a CMOS buffer circuit.

circuit. Meanwhile, the total area increases proportionally, causing a layout consideration.

The following example is the design of a high speed submicron CMOS buffer for a capacitive load of 15 pF.[6] The example has five inverter stages, each of which is larger than the previous one approximately by a factor of e,

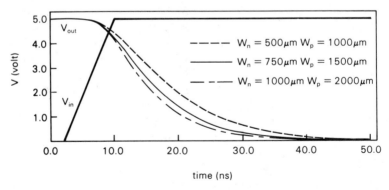

Figure 4.9 Waveforms of a CMOS buffer with various channel widths for the drive transistors (Ref. 5).

and for each stage, the *p*MOS channel width is drawn 1.66 times larger than the corresponding *n*MOS channel width. Using a 0.75 μm CMOS design rule, it was found that the maximum clock frequency designed at 100 MHz needed about 8 mil² (5000 μm²) layout area which is about half of the standard bonding pad size, therefore making it affordable in CMOS buffer design.

For high-speed buffers, dynamic power dissipation is another consideration due to large output load and high frequency operation. Assuming that a CMOS buffer is operated at a standard 5 V supply, the dynamic power per buffer is

$$P_d = CV^2 f = 15pf \times (5 \text{ V})^2 \times 100 \text{ MHz} = 37.5 \text{ mW} \qquad (4.11)$$

For a 40 output chip, the total dynamic power due just to the output buffers amounts to 1.5 W. Reducing output swing from 5 V to 3.3 V or lower using BiCMOS or 0.5 μm CMOS technology can decrease the total power to ≤0.65 W. The BiCMOS technology will be described in Chapter 5, Section 5.10.

4.3 TRANSMISSION GATE AS A SWITCH

Another basic element in a logic circuit is the transmission gate. The gate uses a transistor as a switch to pass signals. It is therefore also called a pass gate or switching transistor. A typical pass transistor in an NMOS circuit is shown in Fig. 4.10a. The capacitance C_L represents loading effect in the circuit. When the clock is on ($\phi = V_{DD}$) and V_{in} is high, the capacitance can be charged toward V_{in}. However, as V_o reaches $V_{DD} - V_t$, the transistor cuts off because V_{gs} of the transistor is now at V_t. So V_o must stay one V_t below V_{DD}. Although subthreshold leakage can eventually charge V_o above $V_{DD} - V_t$, it would take an unreasonably long time.

For all practical applications, V_o can only be charged to $V_{DD} - V_t$. As the clock is turned off, the capacitor is isolated and V_o remains. When the clock is turned on again, and if the V_{in} is low, the capacitor discharges to V_{ss}. The threshold drop associated with the pass transistor causes a problem in passing the full voltage at V_{in} to V_o. This signal degradation is worse if the body effect of the transistor is significant because V_t increases as V_o approaches V_{in}. For example, if V_t (at $V_o = 0$) = 1 V, V_o may not rise to more than 3 V for a 5 V V_{DD} because V_t may rise to 2 V due to the body effect. In NMOS technology, this effect degrades signal propagation and propagation delay. When several pass transistors are connected in series, an inverter is often inserted between two sections of pass transistors to restore logic level and minimize total delay.[1] This problem can be avoided if a CMOS technology is used for implementing the transmission gate.

A CMOS transmission gate consists of a *p*-channel and an *n*-channel transistor with common source and drain connections and separate gates as

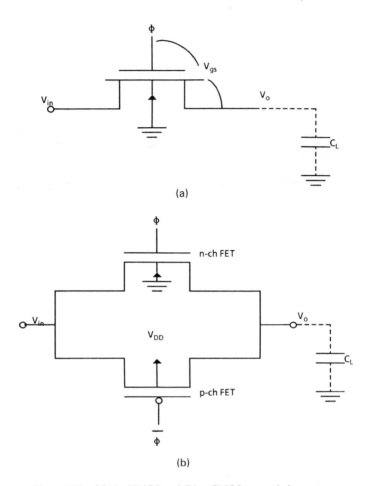

Figure 4.10 (a) An NMOS and (b) a CMOS transmission gate.

shown in Fig. 4.10(b). Furthermore, the complementary clock ($\overline{\phi}$) is applied to the p-channel gate so that both transistors are on or off simultaneously.

While the operation of an n-channel pass transistor has been described, the p-channel transistor operates differently. The transistor can pass V_{DD} from V_{in} to V_o without any degradation. However, it can only discharge V_o to $|V_t|$, at which point the transistor no longer conducts significant current because $|V_{gs}| \leq |V_t|$. So, contrary to an n-channel pass transistor, the p-channel transistor suffers from logic degradation during discharging rather than charging the capacitance. Because the two types of transistors can be paired for a CMOS transmission gate, the p-channel transistor can charge the output to V_{DD} and the n-channel transistor can discharge it to V_{ss}. Thus, logic degradation is observed neither at high nor low levels. The penalty is more transistor counts and additional layout area. In some CMOS circuits, like a pseudo-NMOS design in which only n-channel pass transistors are used, the body effect of the transistors should be minimized. Because the source and drain of a pass transistor can be interchanged during circuit operation, any asymmetric or localized effect caused by hot electron effects can be important in this device. Hot electron effects will be discussed in Chapter 6. More complicated CMOS circuits are described in other books.[3,4]

4.4 CMOS LOGIC GATES

In an NMOS technology, simple logic circuits such as NAND or NOR gates can be implemented by simply adding more transistors in series or in parallel with the driver transistor. In a conventional CMOS circuit however, corresponding p-channel transistors must be added as pull-up load devices. For comparison, Fig. 4.11 (a) and (b) show the schematics of two-input NAND and NOR gates respectively. Both CMOS and NMOS circuits are shown for comparison. Notice that the multiple inputs in the CMOS cases require more transistors and interconnects between the p- and n-channel transistors.

Layouts of the NAND and NOR gates, again for both CMOS and NMOS technologies, are shown in Fig. 4.12. For the NAND gate (AB)$'$, the output is low only when both inputs (A and B) are high; and it is implemented by having the n-channel devices in serial and the p-channel devices in parallel. On the other hand, the NOR gate (A + B)$'$ requires that the output be high if both A and B are low; and it is accomplished by having the n-channel devices in parallel and the p-channel devices in serial.

Fig. 4.13 shows a more complex gate known as the AND-OR-INVERT (AOI) gate (AB + CD)$'$, which is just a NOR of a and b, where a = AB and b = CD. Both NMOS and CMOS representations are shown. a and b are made of series n-devices and parallel p-devices. The complement gate of the AOI is [(A + B)(C + D)]$'$ known as the OR-AND-INVERT (OAI). As shown in Fig. 4.14, it is just a NAND of components consisting of parallel n-devices and serial p-devices. Due to the need of multiple p-devices for load, additional layout area and interconnects are required for CMOS logic circuits.

Alternatively, one can use pseudo NMOS logic or dynamic logic to avoid these disadvantages. In a pseudo NMOS logic, only a single p-channel FET with its gate connected to V_{ss} is used as the load for multiple input logic. This layout is similar to an NMOS logic design with the n-load device changed to a p-channel device. As in NMOS logic, this type of design requires adequate κ ratio and draws static power when the n-channel input is pulled down. The advantage is the elimination of body effect because the pMOS source and substrate are fixed at V_{DD}.

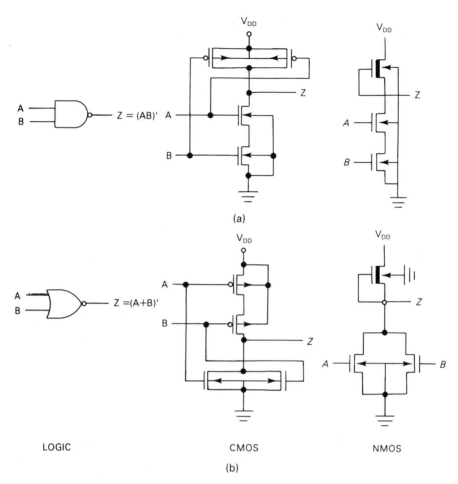

Figure 4.11 (a) NAND and (b) NOR gates implemented by CMOS and NMOS technologies.

4.5 DYNAMIC CMOS LOGIC

Dynamic CMOS logic has also been used for reducing multiple p-channel load devices. As shown in Fig. 4.15, the output node Z is precharged high by the p-channel FET when the clock (ϕ) is low. Z is conditionally discharged to low when ϕ is high, thus turning on the evaluate nMOSFET. Hence the n-channel input logic is evaluated. Cascading this dynamic logic, as shown in Fig. 4.16(a), may cause a problem. During the precharge cycle, output nodes at both stages (N1 and N2) are charged high. Before completely evaluating the first gate, N1 can discharge the output of the following gate

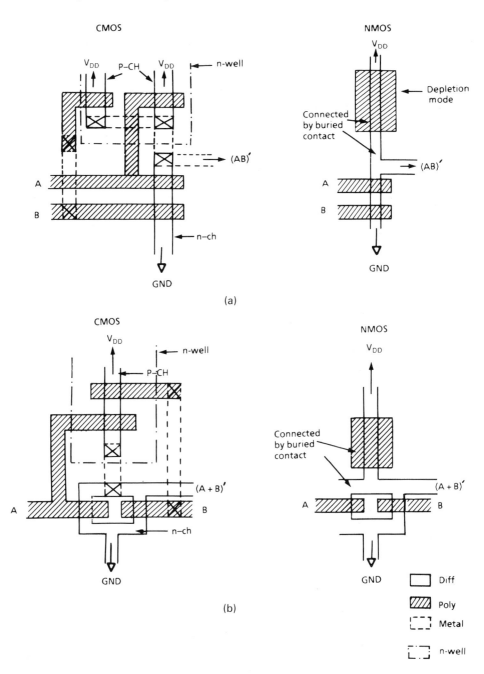

Figure 4.12   (a) NAND and (b) NOR gate layouts using CMOS and NMOS technologies.

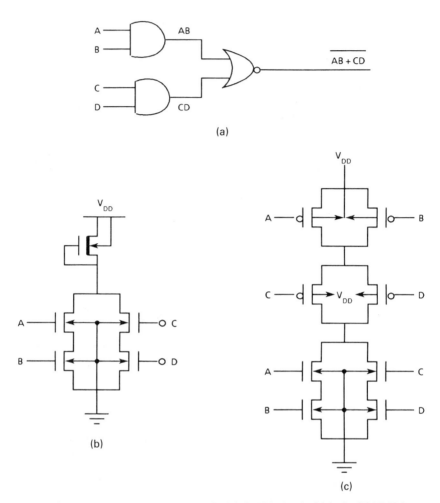

Figure 4.13   An AND-OR-INVERT (AOI) CMOS circuit: (a) logic; (b) NMOS; and (c) CMOS.

(N2 here), resulting in an erroneous state. This problem can be avoided by using multiple clocks or incorporating a CMOS inverter at the output of each logic gate, also known as domino logic. An example of domino logic is shown in Fig. 4.16(b), where the output node *PZ* is inverted to *Z*. During precharge, *PZ* is pulled up high, but *Z* is low. Thus, the input of the subsequent logic gate is always turned off to avoid discharging its output. When the first gate is evaluated, *PZ* conditionally discharges, causing *Z* to go high for evaluating second stage. This way, a single clock can precharge and evaluate all logic gates in a block. Only two *p*-channel transistors are needed for a block of these gates. However, because the inverter buffer can only

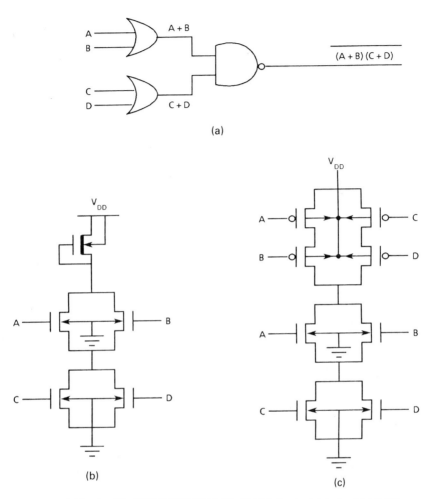

Figure 4.14 An OR-AND-INVERT (OAI) CMOS circuit: (a) logic; (b) NMOS; and (c) CMOS.

make the output go from low to high, only positive logic is possible.   Charge redistribution in this type of dynamic logic may also be a problem.

Cascade voltage switch logic (CVSL) has been proposed to generate any logic expression.   This differential type logic uses a pair of cross-coupled p-channel load devices in connection with two complementary n-channel input logic structures.   As shown in Fig. 4.17(a), when the inputs switch, Q is pulled high or low, and $\overline{Q}$ is then low or high.   Positive feedback causes output to switch.   A dynamic version of the CVSL logic is basically two domino gates operating in a complementary mode as shown in Fig. 4.17(c). Although this kind of logic can generate all logic functions, it suffers from

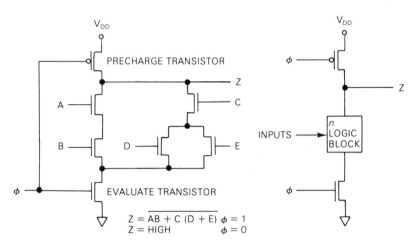

$$Z = \overline{AB + C\,(D + E)}\ \phi = 1$$
$$Z = HIGH \qquad\qquad \phi = 0$$

Figure 4.15 Dynamic CMOS logic (Ref. 3).

the additional routing that is needed for the complement gates, hence extra devices, layout area, and interconnections.

4.6 CMOS RAMS

Random access memories (RAMs) have always been the major driving force of MOS technology because of their requirements for high density, high speed and low power. In addition, because a memory cell is repeated many times for a memory array, a high payoff in shrinking chip area is apparent by using tight design rules and a compact cell layout. The large marketplace associated with RAMs is also another reason for their use. In the past, RAM size has doubled roughly every two-to-four years, and the price has dropped 20%– 30% per year.

There are static and dynamic RAMs. Static memories (SRAMs) use a latch to store data and dynamic memories (DRAMs) store charge on a capacitor. In CMOS, a latch is commonly formed by two cross-coupled inverters, meaning the output of one inverter is connected with the input of the other. In addition, two nMOS pass transistors are tied at output nodes to read and write data. The schematic and layout of a six-transistor CMOS SRAM cell is shown in Fig. 4.18. During the write cycle, data (1 or 0) is placed on the bit line. Hence data is on the bit line. Then, when the word line is selectively turned on, data are written into the cell. To read data out, both bit lines are precharged high. Then, when the word line is selected, one of the bit lines is discharged through the corresponding pull-down transistor.

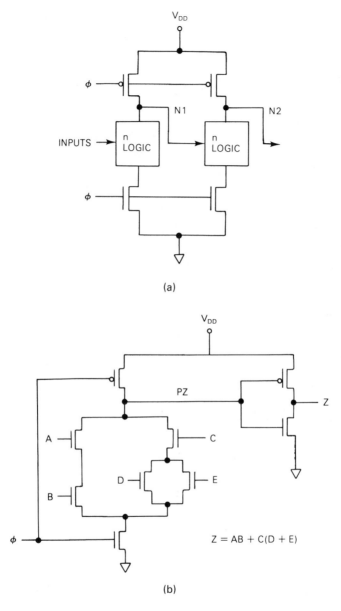

(a)

(b)

Figure 4.16   (a) Cascaded CMOS dynamic gates; and (b) domino CMOS logic (Ref. 3).

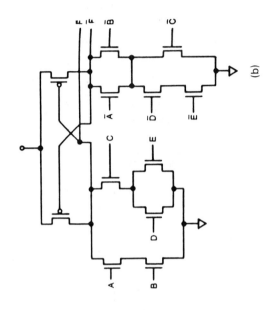

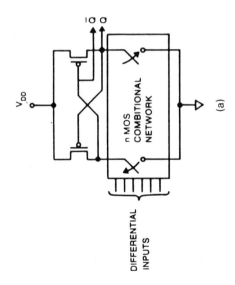

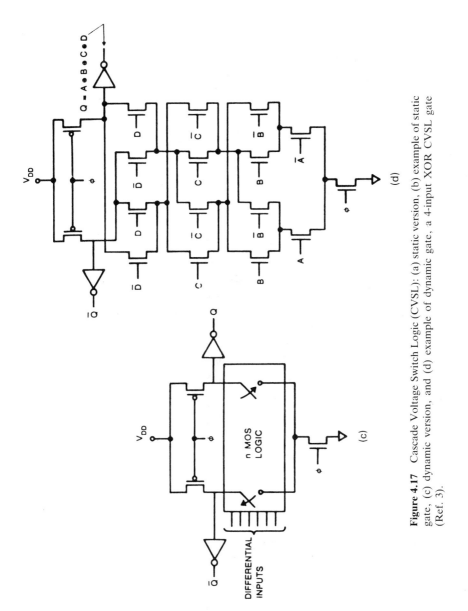

Figure 4.17 Cascade Voltage Switch Logic (CVSL): (a) static version, (b) example of static gate, (c) dynamic version, and (d) example of dynamic gate, a 4-input XOR CVSL gate (Ref. 3).

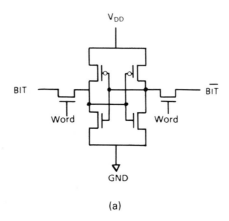

(a)

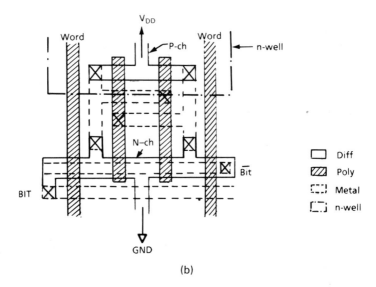

(b)

Figure 4.18 Six transistor CMOS SRAM cell: (a) schematic, and (b) layout.

Once the bit line is pulled low, a differential output signal is generated, activating a sense amplifier as shown in Fig. 4.19. The gate of N3 is biased slightly above the threshold so that N3 is in saturation and acts as a constant current source with a total current I_d and $I_d = I_{d1} + I_{d2}$. Suppose the $\overline{\text{BIT}}$ is the one being pulled slightly lower—I_{d2} will then decrease to $I_{d2} - \Delta$; I_{d1} must be $I_{d1} + \Delta$ to maintain the total I_d constant. Because the two p-channel FETs are load devices, the IR drop across $P2$ is less than that across $P1$, yielding a differential output V_o. Two or more of these stages are often cascaded to provide further amplification. This sensing operation can be fast

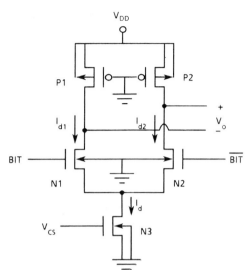

Figure 4.19 A CMOS differential pair for the SRAM sense amplifier.

because one of the BIT lines needs only to be discharged slightly to generate a small differential input. The differential amplifier, described here is also a necessary component for operational amplifiers, which are essential for building analog circuits.

The pMOS load devices in a CMOS SRAM cell provide low power operation, however, at the expense of additional layout area when compared with an NMOS SRAM cell using depletion-mode nMOS FETs or megohm polysilicon resistors for load devices. The poly resistors can be made by a second level of polysilicon on top of the transistors. Enhancements on the six-transistor SRAM cell include dual-port SRAM, which is popular in micro-processors and content addressable memory (CAM). Operations of these RAM cells are described elsewhere.[3,4]

The most often used memory cell in a DRAM is the one-transistor (1-T) cell as shown in Fig. 4.20. Even though other types of DRAM cells use 4- or 3-transistor cells, the 1-T cell configuration is most attractive for high density mega-bit DRAMs. The cell consists of one pass transistor and one storage capacitor, forming very compact cell layout and repeating itself numerous times for high density DRAMs. For the state-of-the-art technology in Mbit DRAMs, roughly one million electrons are used to differentiate high and low states. For CMOS technology, holes can also store data if cell transistors are implemented by pMOSFETs. This scheme offers the advantage of reduced substrate current because the hole impact ionization rate is much lower than that of electrons. Details of this advantage will be discussed in the next chapter.

To write a cell, data is first placed on a bit line, then transferred into the capacitor by the pass transistor. To read a cell, a bit line is precharged,

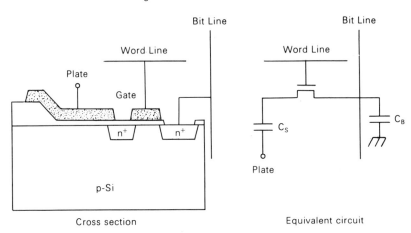

Single Transistor DRAM Cell

Cross section   Equivalent circuit

Figure 4.20 One-transistor DRAM cell structure and equivalent circuit.

then the word line is raised to pass charges from the storage capacitor to the bit line. Charge redistribution changes the voltage on the bit line of the selected cell. This voltage, when compared with the bit-line voltage from a dummy cell, results in a differential signal which in turn activates a sense amplifier. The sense amplifier amplifies the voltage difference to a full logic swing. In this read operation, the information in the selected cell is destroyed. The information must be written back immediately. Due to the limitation of gain bandwidth product in the sense circuit, the stored charge level is critical to DRAM design. On the other hand, cell area is also precious because of the limited chip size for megabit memories. Innovative technologies such as the high-C cell, stacked capacitor, and trench capacitor have been developed to increase stored charge per unit cell area.

Another constraint in signal charge is the signal degradation caused by alpha particles. These high-energy particles penetrate the silicon substrate for roughly 25 μm, generating a track of electron-hole pairs. Minority carriers, when collected by memory cells, can degrade signal levels, thereby resulting in soft errors. Low collection efficiency is therefore important in DRAM technology. As mentioned in Chapter 1, collection efficiency decreases drastically when DRAM cells are made in the well of a CMOS configuration. Figure 4.21(a) shows that for a given number of electron-hole pairs generated by alpha particle radiation, the probability of collecting minority carriers in DRAM memory cells is reduced by orders of magnitude when an n-well is used in a p-substrate.[7] The reverse biased n-well/p-substrate junction offers a retarded field (Fig. 4.21b) to stop the minority carriers (holes in this case) from entering into the arrays built in the well.

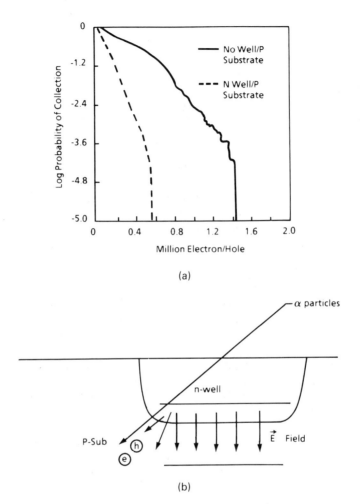

Figure 4.21   Charge collection efficiency in DRAM: (a) probability of collection vs. electron/hole pairs generated, and (b) the schematic of a CMOS cross-section showing the associated retarding electric field (Ref. 5, © 1983 IEEE).

REFERENCES

1. C. Mead and L. Conway, *Introduction to VLSI Systems,* Addison-Wesley, 1980.

2. D. A. Hodges and H. G. Jackson, *Analysis and Design of Digital Integrated Circuits,* McGraw Hill Book Co., 1983.

3. N. Weste and K. Eshraghian, *Principles of VLSI Design,* Addison-Wesley, 1985.

4. L. A. Glasser and D. W. Dobberpuhl, *The Design and Analysis of VLSI Circuits,* Addison-Wesley Co., 1985.

5. M. Annaratone, *Digital CMOS Circuit Design*, Kluwer Academic Publishers, p. 7.21, 1986.

6. J. Y. Chen, "Scaling CMOS to submicron design rules for VLSI," *VLSI Design Magazine*, p. 78, July 1984.

7. P. Madland, J. Schutz, R. Green and R. King, "CMOS vs. NMOS comparisons in dynamic RAM design," in the Proc. of International Conf. on Computer Design: VLSI in Computer, p. 379, 1983.

EXERCISES

1. Prove that CMOS inverter delay time is minimal if $L_p = L_n$ and $W_p = \sqrt{\mu_n/\mu_p}W_n$.

2. A CMOS inverter has the following dimensions: $t_{ox} = 200$ Å; $t_{poly} = 3000$ Å; $L_n = L_p = 1$ μm; $W_p = 2W_n = 10$ μm; $\mu_n = 2\mu_p = 600$ cm²/V-sec; $l_{ov} = 0.1$ μm; and, $V_{tn} = -V_{tp} = 1$ V. There are four 2 μm-wide metal lines at the output node at a length of 250 μm each. They are 1 μm apart on a field oxide with 1 μm thickness. Estimate the Miller capacitance (C_m) and stray capacitance (C_s), then calculate total delay time for charging and discharging the output node.

3. If the *p*-channel device in the CMOS inverter described in Exercise 2 has a leakage current of 1 μA when the gate is off, calculate the output voltage when its input is at 5 V.

4. For a CMOS buffer made of an inverter chain with load capacitance of C_L and internal inverter input capacitance of C_g, prove that the total delay is minimal if each inverter is larger than the preceding one by a factor of e, where e is the base of natural logarithms, and the number of inverters is $\ln(C_L/C_g)$.

5. An NMOS pass gate is connected by a 5 V clock voltage, a 5 V input voltage and the output load is 5 pF. The transistor characteristics are: $V_t = 0.5$ V; $W/L = 10/1$; $t_{ox} = 200$ Å; $\mu_n = 500$ cm²/V-sec; and, the body effect coefficient (γ) is 0.5 V$^{1/2}$. What voltage level can the output reach in a reasonable charging time? What is the estimated charging time?

6. For the pass gate in Exercise 4, if the subthreshold swing of this transistor is 100 mV/dec and threshold voltage is defined at 1 μA, and $\gamma \cong 0$, estimate the additional time to charge the output to 5 V.

7. What are the advantages and disadvantages of a domino CMOS logic in comparison with a static CMOS logic? Is *n*- or *p*-well a better technology to implement the domino logic?

5

CMOS PROCESS TECHNOLOGY

This chapter discusses CMOS process technology, emphasizing process architectures, meaning the masking sequences and overall process integrations, rather than individual process modules. First introduced is the relationship between a circuit layout generated by designers and a device cross-section at various key processing steps. Then various CMOS process architectures are described: p-well, n-well, twin-tub and retrograde well processes. Comparisons will be made between these different processes and their impacts on circuit performance will be discussed. Reasons for selecting a particular process architecture or process module will be presented in this chapter, and further explained in later chapters in conjunction with device design. CMOS devices built on sapphire and other insulators are also described. Finally, bipolar and CMOS integration, called BiCMOS, is discussed extensively. Bipolar device characteristics, BiCMOS operation, BiCMOS process technology and BiCMOS applications are all included in the discussion.

5.1 PROCESS ARCHITECTURE

A standard IC process involves transferring an IC layout to Si wafers. This process is normally done by printing IC layers from a set of masks through photolithography followed by subsequent processing such as etching, implantation, diffusion, etc. The sequence of applying the photolithographic masks is important for the fabrication of devices and ICs. In general, a bottom-up approach (meaning the layers at the bottom are formed first) is used because the layers are stacked during IC processing. However, options exist in permutating the masking sequence and even the total number of masks. Trade-off commonly occurs between the number of masks and process complexity.

The art of defining mask count and sequence and overall process integration will hereby be referred to as process architecture. A process architect is responsible for fabricating ICs by integrating the necessary process modules in the appropriate order. Mask counts, process simplicity and device performance are important considerations in establishing a good process architecture.

To establish a process architecture, the correspondence between the IC layout and device cross-section at any process step must be fully understood. A CAD program, SIMPL-2 (SIMulated Profiles from the Layout—Version 2), has been developed[1] for this purpose. Figure 5.1 shows an example for a CMOS process. The layout of a CMOS inverter and its corresponding cross-sections are generated by the SIMPL-2 program. The bold horizontal line on the layout is the user-specified "cut-line" along which a cross-sectional view is generated.

A CMOS process architecture has to provide both n- and p-channel transistors on a Si wafer. As a result, various choices exist in defining the process architecture. Forming a p-well in an n-type substrate is one option, or an n-well can be created in a p-substrate. Other alternatives are twin-tub,[2,3] retrograde-well,[4,5] and quad-well technology.[6] Which one should be chosen for VLSI has been a controversial subject.

One reason for the CMOS controversy is that two device types used in the technology perform differently, and at least one of them must be located in a well. As described in Chapter 2, an n-channel transistor delivers more drain current because electrons move faster than holes; however, unwanted substrate current in an n-channel device is also higher due to higher impact ionization. Figure 5.2 shows these two fundamental differences between n- and p-channel devices.[7] The devices in the well suffer from higher junction capacitance and stronger body effect. Substrate current in the well is also harder to collect. A device technologist may think that if one type of device has to suffer from CMOS integration, it ought to be the better device (n-channel in this case) to balance the performance between the two devices. A circuit designer, on the other hand, may want the better device to be maintained or further optimized and choose to avoid using the other type. One well-known example is the domino logic (described in Chapter 4), which uses only one p-channel device for several n-channel transistors. Whether CMOS circuits should be made truly complementary depends on the per-

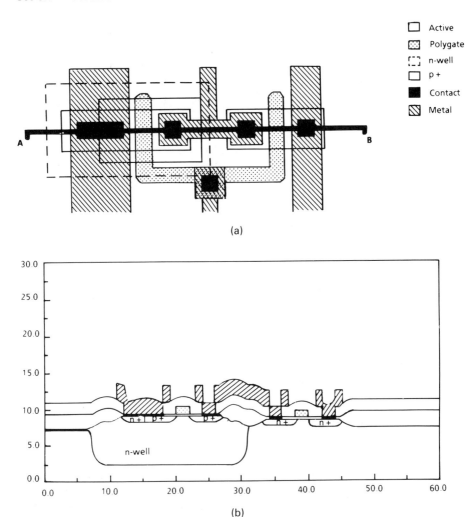

Figure 5.1   (a) A CMOS inverter layout; and (b) cross-section simulated by SIMPL-2 (Ref. 1).

formance difference between the two device types.   The ratio of the p-channel saturation current to the n-channel saturation current increases from 1/4 to 1/2 when the devices are scaled to 1 μm [Fig. 5.2a] because electron and hole velocities start saturating to a common asymptote.   Below that, the p-channel current approaches the n-channel current.[8,9]   Thus the twin-tub approach, which discriminates neither n- nor p-channel devices, would be attractive. In the following sections, we will discuss all of these options and compare their pros and cons.

FUNDAMENTAL DIFFERENCE BETWEEN n- AND
p-CHANNEL FET's

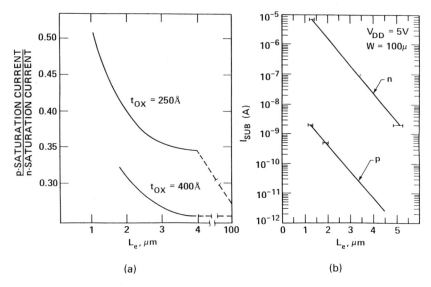

(a)   (b)

Figure 5.5 Cross-sections, masking levels and associated processes of an *n*-well CMOS front-end process.

5.2 *P*-WELL PROCESS

Conventional CMOS was realized by putting *n*-channel transistors in a *p*-well formed by diffusing boron atoms into an *n*-type substrate. *P*-channel devices were made outside the well, in the *n*-substrate. Figure 5.3 shows the layout masks and the corresponding device cross-sections of a typical *p*-well CMOS process. The first mask defines *p*-well regions by opening windows in an oxide masking layer. Boron atoms are introduced into these windows by a shallow ion implantation followed by a high-temperature diffusion. A *p*-type well with a depth of a few microns is then formed (Fig. 5.3a). A layer of nitride is then deposited and patterned to cover the active areas in which transistors are to be built. LOCOS (LOCal Oxidation of Si) is used to grow a thick (0.5–1.0 μm) field oxide among the active areas (Fig. 5.3b). After etching away the nitride layer, a thin gate oxide (200–500 Å) is grown on the active areas as the gate oxide. A layer of polysilicon material is deposited and doped to *n*-type by phosphorus. The poly-layer is then patterned and reactive-ion-etched to form MOS gates for both *n*- and *p*-channel FETs (Fig. 5.3c). To form source/drain (S/D) regions, a layer of photoresist is patterned to have large windows opened on one type of FETs, e.g., *p*-channel FETs.

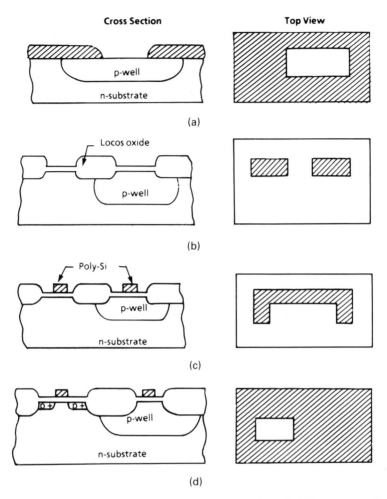

Cross Section   **Top View**

(a)

Locos oxide

p-well

(b)

Poly-Si

p-well

n-substrate

(c)

p-well

n-substrate

(d)

Figure 5.3 Cross-sections and top views of a typical *p*-well CMOS process at all mask levels: (a) *p*-well mask; (b) active area mask; (c) poly gate mask; (d) p^+ mask; (e) n^+ mask; (f) contact mask; and (g) metal mask.

Boron ions are implanted into the *p*-channel active areas forming p^+ source and drains (Fig. 5.3d).   Poly-gates in the active areas block boron ions going to the channel region, resulting in p^+ self-aligned to the poly-gates.

A similar process is applied to form n^+ source and drains for *n*-channel FETs by implanting arsenic ions through an n^+ mask (Fig. 5.3e), which is basically the complement of the p^+ mask.   Following the stripping of the photoresist layer, a layer of oxide is Chemically Vapor Deposited (CVD) and contact holes are patterned and etched on gates, p^+ region, and n^+ regions to make electric contacts (Fig. 5.3f).   A layer of metal (normally aluminum)

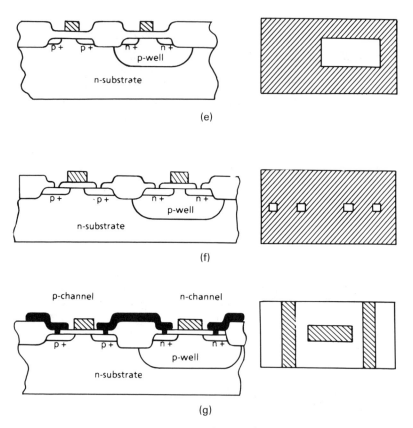

(e)

(f)

(g)

Figure 5.3 *Continued.*

is deposited and an interconnect pattern is defined. A low-temperature anneal (450°C) usually completes the CMOS process (Fig. 5.3g).

The entire process consists of seven masks in the sequence of: p-well, active area, poly, p^+, n^+, contact and metal. This sequence describes only a basic CMOS process. State-of-the-art CMOS technology often requires threshold-adjust and punchthrough-stop implants for the reduction of short-channel effects. One or two more masks are commonly added after the definition of the active area. Double-level metal is another add-on for a VLSI process. A layer of CVD oxide is deposited on top of the first-level metal, then Via holes are patterned and etched. A second-level metal is then deposited, patterned, and annealed. It is also common industry practice to deposit a passivation layer and open windows over metal pad regions for wire bonding. These additional processes add up to a CMOS process with approximately a dozen masks. For ≤ 1 μm CMOS processes, a lightly doped drain structure is often included to reduce electrical field at the expense of more processing steps.   Details of this process will be discussed in Chapter 6.

5.3 *N*-WELL PROCESS

Recent development has resulted in an *n*-well technology[7,10] with a structure identical to Fig. 5.3g if *n*- and *p*-labels are interchanged in the figure.   The idea is to make *n*-channel FETs in a low-resistivity *p*-type substrate rather than a more heavily doped *p*-well.   *N*-channel devices formed by this CMOS process are equivalent to FETs produced by NMOS technology.   Device design and process architecture for NMOS technology is therefore transferable to this CMOS process.

Process architecture is similar to *p*-well architecture with the exception that phosphorus atoms are used to form *n*-wells in a lightly doped *p*-substrate. In addition, the field region outside the *n*-wells should be doped to enhance isolation for *n*-channel FETs.   This additional doping can be done with boron implantation before LOCOS.[11]   The boron dose should be sufficiently high to offer high field threshold for *n*-channel transistors, but still low enough so that the boron concentration is fully compensated by the *n*-well dopants under the field oxide.   Consequently, a reasonable field threshold for *p*-channel devices can be maintained.   The finished CMOS device structure is shown in Fig. 5.4.

Another alternative is to implant boron atoms only in the *n*-channel field region to avoid boron/phosphorus compensation.   However, this process would require an extra mask and slightly larger layout area for maintaining the separation between the *n*-well and the *p*-type channel stop.   The process sequence for this architecture is shown in Fig. 5.5.   Notice that the *n*-channel field implant mask overlaps the actual *n*-well edge (Fig. 5.5b), which extends beyond the *n*-well mask edge due to lateral diffusion.   Shallow implants for adjusting *n*- and *p*-channel threshold voltages are also included.

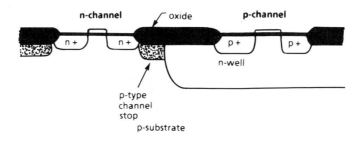

Figure 5.4   Cross-section of a finished *n*-well CMOS structure with self-aligned *p*-type channel stop.

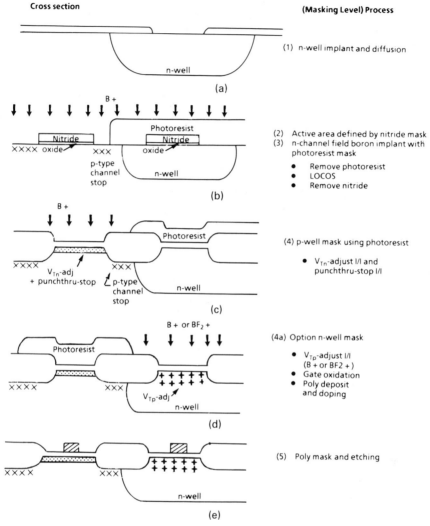

Figure 5.5 Cross-sections, masking levels and associated processes of an *n*-well CMOS front-end process.

As shown in Fig. 5.5(c), a *p*-well mask is added when boron ions (B$^+$) are implanted to adjust the *n*-channel threshold (V_{Tn}). The same implant or a second B$^+$ ion implant (I/I) increases the channel doping concentration so that punchthrough is prevented. This second implantation is often necessary for short-channel *n*MOS transistors built in a lightly doped substrate. A B$^+$ or BF$_2^+$ implant is also needed to adjust the *p*-channel threshold (V_{Tp}) with or without a mask (the optional *n*-well mask in Fig. 5.5d). If the V_{Tp}-adjust implant dose is less than what is needed for the V_{Tn} to adjust, the V_{Tp}-adjust implant can be made to both *n*- and *p*-channel areas with no mask needed.

The extra dose needed for the *n*-channel can be fixed at the V_{Tn}-adjust ion implant. The remaining process steps follow conventional MOS process.

5.4 *P*-WELL VERSUS *N*-WELL

Even though no process clearly wins today in selecting a technology, it is useful to know the pros and cons of the two distinct well configurations and their impacts on circuit applications.

Present CMOS technology offers 1–2 μm design rules. At these dimensions, *n*-channel devices, when compared with *p*-channel devices, provide about twice the driving current but almost four orders of magnitude more substrate current.[7] These fundamental differences arise from carrier mobility and the impact ionization rate (described in Chapter 2), and they affect technology selection in many ways. Another fundamental limit is that the doping concentration in the well has to be higher than in the starting substrate, thus resulting in higher junction capacitance and more body effect for devices made in the well.

Material-related issues and process considerations serve as practical constraints. For example, epitaxy material types that have been used to reduce latch-up susceptibility have different implications. It has been shown that *p*-type epi grown on p^+ substrate provides a longer (msec) minority lifetime than the case of *n*-epi on n^+ substrate, due to better intrinsic gettering.[12] But, out-diffusion of *p*-on-p^+ is much more severe than that of *n*-on-n^+ because boron diffuses much faster than arsenic or antimony. From this point of view, DRAMs or other dynamic circuits should be built on *p*-on-p^+ epi (meaning *n*-well technology), whereas static circuits ought to be made in *n*-on-n^+ epi (*p*-well) for sharper epi interface, hence better latch-up protection.

It is also known that, during oxidation, boron segregates into oxide but phosphorus piles up at the silicon surface. This phenomenon and the fact that the oxide fixed charge is normally positive make the field region among *n*-channel transistors more sensitive to inversion problems. While a *p*-well process can use the well itself as an *n*-channel channel stop, an *n*-well process has the burden to produce separate *p*-type channel stops for *n*-channel devices. Moreover, if ion implantation is used rather than diffusion to form a well (a so-called retrograde-well), it is easier to form a *p*-well than an *n*-well because boron ions penetrate deeper than arsenic or phosphorus ions for a given implant energy.

The choice of well type depends highly on circuit applications. For *n*-MOS rich circuits, such as domino logic or cascade voltage switch logic,[13] *n*-well technology, which allows *n*-channel transistors to be built in the substrate, should be chosen. On the other hand, the *p*-well approach may be a better choice for pure static logic to balance the performance of the two device types. *P*-well also favors devices that require an isolated *p*-region, e.g., *n*-channel FETs for analog input.

In the case of RAMs, if alpha-particle induced SER (Soft Error Rate) is a major limitation in RAM scaling, the RAM cells ought to be made inside a well. For DRAMs, *p*-channel arrays in an *n*-well might be suitable because *p*-channel devices have low substrate current whereas the high substrate current in *n*-channel devices can be easily sunk when they are in substrate. The *p*-channel speed disadvantage does not degrade DRAM performance greatly because DRAM sensing is limited primarily by the amount of charge stored.[14] SRAM sensing is, however, different and it depends on current provided by the SRAM cell. Hence, high gain *n*-channel transistors are more desirable for pass gates and drivers in a cell. A cell should be put in a *p*-well for low soft error rate as described in Chapter 1. For high voltage applications, such as EPROM and EEPROM, an *n*-well is appropriate because sinking *n*-channel substrate current is critical.

5.5 TWIN-TUB PROCESS

The twin-tub approach forms two separate wells for *n*- and *p*-channel transistors in a lightly doped substrate. The complete device structure is shown in Fig. 5.6 in which the substrate can be either *n*- or *p*-type. The original claim for this structure was that doping profiles in each well could be set independently; hence both device types would be optimized.[2] In a single-well scheme, the substrate doping concentration must be an order of magnitude lower than the well concentration but meanwhile, must be high enough to prevent punchthrough for the devices made in the substrate. This claim is not always true, because state-of-the-art MOS technology uses a shallow implant to prevent punchthrough without raising the entire substrate doping

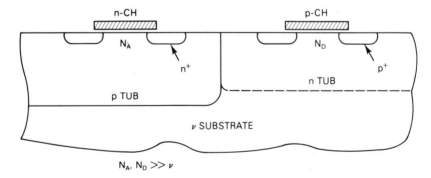

$N_A, N_D \gg \nu$

Figure 5.6 Twin-tub CMOS structure (Ref. 2).

concentration.   In terms of lower junction capacitance and less body effect, this method actually produces better devices than those attainable with the twin-tub approach.

ɪne major advantage of the twin-tub approach is the flexibility of selecting substrate type (*n* or *p*) with no effects oɔ transistor performance; the latchup behavior however, will not be identical.   This flexibility may be important in implementing designs with different applications.   In addition, self-aligned channel stops can be easily implemented with this approach. Consequently, spacing between an *n*- and a *p*-channel device can be reduced for high density circuits.

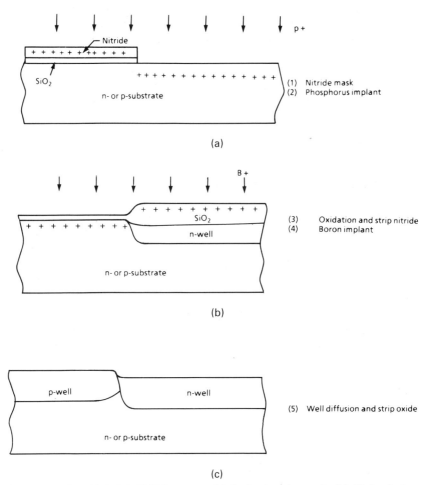

(a)

(b)

(c)

Figure 5.7   Twin-tub CMOS process. (a) P^+ implant for *n* tub; (b) B^+ implant for *p* tub; and (c) well diffusion for both tubs (Ref. 2, © 1980, IEEE).

Figure 5.7 shows the process sequence at the front-end of the twin-tub process. A layer of patterned nitride is used to mask a phosphorus implant for *n*-tubs. Subsequent to the growth of a masking oxide, the nitride layer is removed and the oxide is then used as a mask during boron implantation for *p*-tubs. With this method, *n*- and *p*-tubs are formed using only one mask. Moreover, the two tubs are mutually self-aligned. The remaining process follows active area definition using LOCOS and similar process steps described for *p*-well process architecture.

As CMOS technology advances to submicron dimensions, the twin-tub approach may become more attractive for the following reasons. Because the two device types perform similarly in the half-micron regime, it makes sense to provide symmetric *n*- and *p*-channel devices.[15] Because the doping concentration will be scaled up at these dimensions anyway, whether devices are made in the well or substrate makes only a marginal difference. Submicron technologies such as trench isolation and epi substrate work well with the twin-tub approach. For example, trench sidewall inversion is less likely when both sidewalls are butted against highly doped wells. This problem will be discussed in Chapter 7. Moreover, when epi is used, this approach offers greater flexibility in choosing n-on-n^+ or p-on-p^+, and even n-on-p^+ or p-on-n^+ if BiCMOS (bipolar/CMOS) chips are implemented.

5.6 RETROGRADE-WELL PROCESS

Conventional wells are formed by diffusion, which is an isotropic process, meaning impurity atoms diffuse laterally as well as vertically. Lateral diffusion takes up Si area resulting in poorer packing density. High-energy ion implantation, when used for well formation, provides minimal lateral spread because of the anisotropic nature of the implantation process. In a retrograde-well process architecture, wells are formed after active area definition and LOCOS. As a result, the lateral spread of the well is further minimized because the high-temperature LOCOS process is done prior to well formation. Figure 5.8 shows the cross-section of an implanted *p*-well in comparison with a conventional diffused well.[4] In this example, the implanted *p*-well is formed by boron implantation at 400–600 KeV followed by a brief 30 min. anneal at 1000°C instead of a 20 hr. drive-in at 1100°C used for conventional well formation. Notice that the lateral spread associated with the *p*-well is greatly reduced. This decrease in spread has led a reduction of p^+-to-n^+ spacing from 12 μm to 9 μm. As a result, a 4K-bit CMOS SRAM was shrunk by 25% in all linear dimensions using this process.[4]

Unlike a diffused profile in which peak concentration is always at the Si surface, the peak of the implanted profile is buried at a certain depth (depending on the implant energy) inside the Si substrate and the impurity concentration decreases as it approaches the Si surface. This type of profile

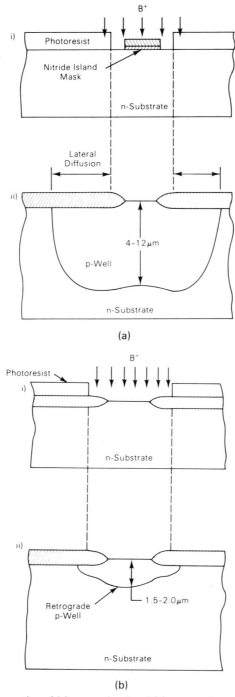

Figure 5.8   Formation of (a) conventional and (b) retrograde p-well CMOS structures (Ref. 4, © 1981, IEEE).

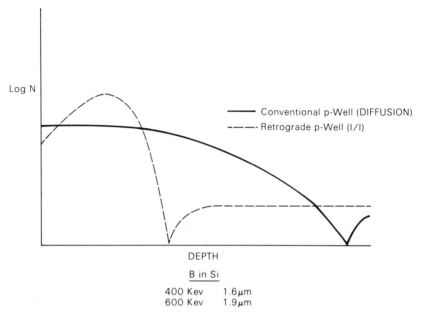

Figure 5.9 Comparison of conventional and retrograde p-well profiles (Ref. 4, © 1981, IEEE).

is called retrograde profile, and the implanted well is often referred to as the retrograde well. Comparison of conventional and implanted p-well profiles are shown in Fig. 5.9. An n-type retrograde-well process has also been proposed using an even higher implant energy: 700 KeV phosphorus.[16] As shown in Fig. 5.10, the process involves lifting-off a layer of evaporated Si film and using a $Si/Si_3N_4/SiO_2$ multi-layer film as a mask. As a result, the p-type channel stop is self-aligned to the n-well, but at the expense of process complexity. A simpler retrograde n-well process was recently demonstrated.[11]

Figure 5.11 shows the front end of the process. Similar to a standard nMOS process, nitride is used as a mask for blanket field boron implant and LOCOS. Next, a resist mask is used to implant phosphorus ions for n-well formation. Because the phosphorus concentration for the n-well portion underneath the field oxide is an order of magnitude higher than the boron concentration there, the resultant net impurity is n-type with 10^{17} cm^{-3} concentration. Quadruple-well structure[6] uses two very shallow wells in the field region as channel stops in addition to the two relatively deep wells for active transistors. Both deep wells are retrograde. Figure 5.12 shows the process architecture in which liftoff is used. However, liftoff can be eliminated by using an extra mask.

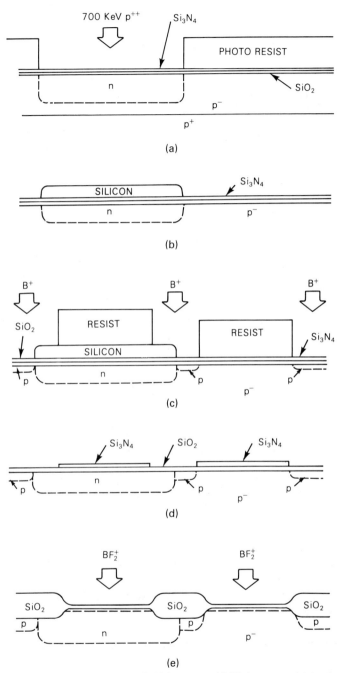

Figure 5.10 A retrograde n-well CMOS process. (a) High energy p^{++} implant to form n-well with a resist mask; (b) the same resist is used to lift off Si; (c) the remaining Si and another resist mask are used to implant B^+ for a p-type channel stop; (d) transfer of the resist mask to the underlying nitride; and (e) LOCOS and strip resist, then BF_2^+ for channel implants (Ref. 16, © 1985, IEEE).

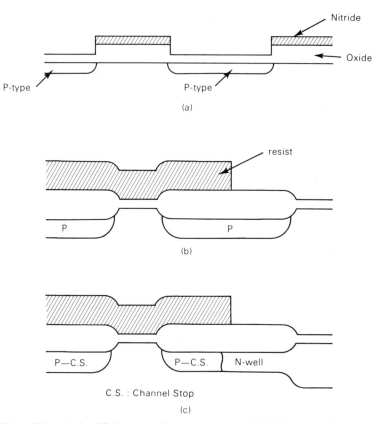

Figure 5.11 A simplified retrograde n-well process. (a) Nitride mask for active area and blanket B^+ implant for p-type channel stop; (b) photoresist mask for n-well; and (c) high energy P^+ implant and any additional shallow implants in the n-well (Ref. 11, © 1986 IEEE).

The major benefit of retrograde-well processes is high packing density through the reduction of p^+-to-n^+ spacing. This approach is desirable for radiation-hard applications because of the need for high field threshold voltages at a small isolation spacing and it is also scalable for half-micron technology. Other advantages are:

(1) Providing a retarded electrical field that reduces the current gain of the vertical bipolar transistor.

(2) Increasing conductivity at the bottom of the well, which decreases well resistance, hence enhancing latchup resistance. It also increases breakdown voltage of vertical punchthrough between the drain in the well and the substrate.

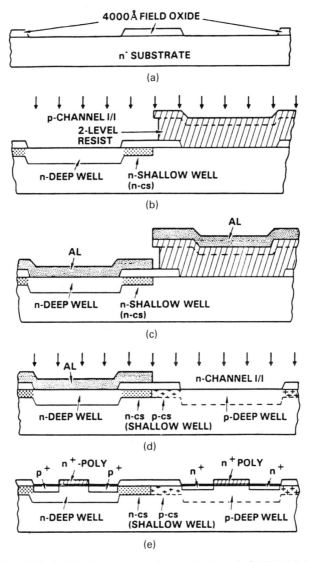

Figure 5.12 Fabrication sequence for quadruple well CMOS. (a) Define active area by oxide windows; (b) a thick, preferable 2-level photoresist mask for *n*-well implant; (c) evaporate and lift off a layer of Al; (d) implant B^+ for *p*-wells; and (e) strip Al and follow standard processes to complete the CMOS structure (Ref. 6, © 1984 IEEE).

(3) Reducing junction capacitance and body effect if the implant energy is sufficiently high (approaching 1 MeV) to move the highly doped region away from the channel.

(4) Enhancing latchup immunity, because its low thermal cycle is more compatible with a shallow epitaxial layer, which is crucial for latchup resistance.

These benefits will be discussed in Chapter 8. Moreover, for BiCMOS applications, the highly conductive layer near the bottom of the retrograde well can also be used as a buried layer if bipolar devices are made in the well.

Concerns for retrograde-well processes are the need for thick photoresist as a masking material for high energy implant, annealability of Si damage caused by high energy ions, and other practical considerations for high-volume production. Although the retrograde-well technology has been shown in laboratories, it should be mentioned that manufacturability has not been established in production due to the need of high-throughput implanters operating at ion energies above 400 KeV.

5.7 CHOICE OF PROCESS ARCHITECTURES

In addition to considering process simplicity and device performance, the choice of a particular CMOS process architecture strongly depends on circuit applications. As described above, a p-well process may be desirable for SRAM design if alpha particle induced soft error is the main concern. On the other hand, for most logic circuits designed with Domino logic, an n-well approach is more suitable. The relationship between process choice and circuit design is summarized in Table 5.1. For analog circuits, high gain nMOS devices are preferred over pMOS devices. In addition, the substrate is often connected to the source to avoid body effect; therefore, isolated p-wells are required.

Process choice also relies on the historical development of a Si house; for example, a well-established NMOS facility may have chosen the n-well approach because all of the n-channel processes are portable, whereas an established CMOS manufacturer may maintain its original p-well technology so that existing designs can be used. Scaling of CMOS to half-micron dimensions may make twin-tub or retrograde-well technologies more attractive for the reasons previously mentioned.

5.8 SOS TECHNOLOGY

All of the process architectures described so far are based on bulk CMOS technology from which most CMOS ICs are fabricated. However, another CMOS technology involves making circuits on a sapphire instead of a silicon

TABLE 5.1 Relationships Between CMOS Technology Choice and Circuit Design

Technology Consideration	Impact on Design	p-Well	n-Well	Twin-Tub
Optimizing nMOS performance	nMOS rich, dynamic circuits		√	
Balancing nMOS and pMOS performance	Static	√		
Sinking substrate current caused by impact ionization	DRAM and E²PROM		√	
Collecting α-particle-induced carriers	RAM cells in a well	√	√	√
Using low l_{sub} FETs for memory arrays	DRAM cells		√	
Using high-gain drivers and pass gates	SRAM cells	√		
Eliminating body effect	Analog input	√		
Switching substrate type	Design flexibility			√
Scaling down to submicron	Future VLSI			√
Using epi: n/n^+	Latch-up-free	√		
p/p^+	DRAM		√	
Using trench isolation		√		√
Using retrograde well	High density	√		
Integrating bipolar on CMOS chip	BiCMOS		√	√
Using well-established CMOS technology	Existing designs	√		
Modifying existing nMOS process	nMOS compatible		√	

substrate. Sapphire is an insulating material with its lattice sufficiently matched to silicon so that silicon can be epitaxially grown on the sapphire substrate. This structure, referred to as silicon on sapphire (SOS), offers some unique advantages for CMOS technology, especially for military and aerospace applications.

In a CMOS/SOS process, sapphire wafers with a thin (0.3–0.5 μm) Si epi layer are commonly used as a starting material. Most IC facilities purchase them from SOS wafer suppliers. SOS wafers in general are smaller in size (2–3 inches in diameter) and cost much more than bulk Si wafers. Crystal defects also exist at or near the Si/sapphire interface and techniques such as solid-phase epitaxial re-growth[17] have been attempted to improve SOS material quality before IC processing.

As shown in Fig. 5.13, a CMOS/SOS process normally starts with making Si islands by etching grooves down to the sapphire substrate. These islands are made on the insulating substrate and separated by air. Using ion implantation with photoresist masks, p- and n-type islands are formed for n-

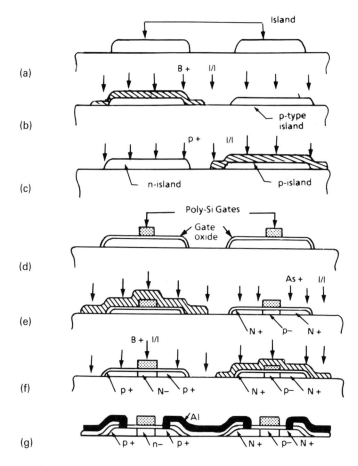

Figure 5.13 CMOS/SOS island process. (a) Island definition by etching Si; (b) B^+ I/I for nMOS; (c) P^+ I/I for pMOS; (d) Poly gate definition; (e) As$^+$ I/I for nMOS S/D; (f) B^+ I/I for pMOS S/D; and (g) contacts and metal to complete the process.

and *p*-channel MOSFETs, respectively.   Subsequent process steps are similar
to the bulk CMOS process discussed in Sec. 5.2.   An obvious advantage of
this process is that MOSFETs are nicely isolated with a minimum of isolation
spacing.   However, the corresponding topography has made step coverage
of poly or metal lines over the island edges difficult.   An alternative is to use
a nitride mask for partial Si etching (Fig. 5.14a) and recessed LOCOS.   This
process results in a planar structure at the expense of the LOCOS-induced
bird's beak.   The remaining process steps again are identical to a standard
CMOS process.   Both process architectures require a minimum of eight masks
consisting of: active area, *p*-island, *n*-island, poly-gate, n^+, p^+, contacts and
metal.

In addition to the advantage of device isolation provided by the insulating substrate, SOS offers several other benefits.   First, latchup does not
exist because there is no *pnpn* path in SOS.   Second, because its S/D regions
are located directly on the sapphire substrate, junction capacitance and soft
error are minimized.   Third, with the thick (300 μm) insulating substrate,
metal to substrate capacitance is low.   Finally, the corresponding process can
be simpler because neither well nor channel stop is needed.

Several disadvantages are associated with this technology in addition to
the wafer size and cost previously mentioned.   Crystal defects at the Si/
sapphire interface always produce lower electron mobility in SOS, even though
its hole mobility is comparable to that of bulk Si.   Moreover, higher leakage
current is normally observed in SOS due to the parasitic back-channel FET

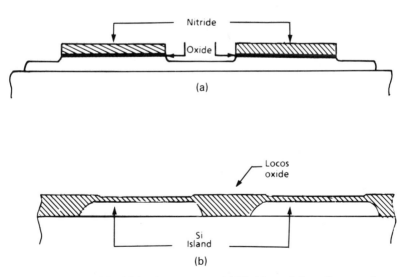

Figure 5.14   CMOS/SOS isoplanar process. (a) Nitride mask for active area, then
etching half of the Si layer away; and (b) LOCOS oxidation to form recessed (i.e.,
planar) field oxide.

formed at the Si/sapphire interface.[18] SOS island sidewalls are another source of leakage. These leakage problems will be discussed in Chapter 7. Lower mobility degrades circuit speed and higher leakage reduces noise margin and increases power dissipation. New techniques have been sought to reduce SOS defect density; however, difficulty increases with the thinner SOS films that are desirable for making short-channel transistors.

INTERCONNECT CAPACITANCE
vs LINE/SPACE SIZE

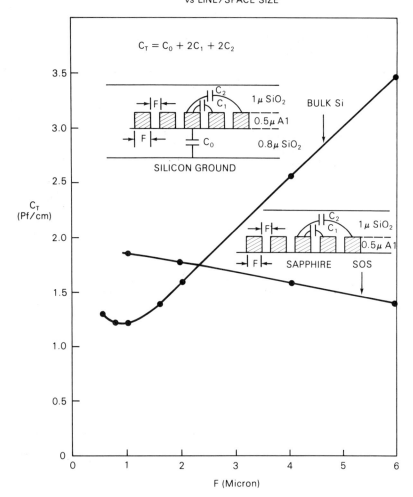

Figure 5.15 Interconnect capacitance vs. line/space size for bulk and SOS technologies. C_o is the metal to substrate capacitance; C_1 and C_2 are the metal line to nearest and to next nearest neighboring metal line capacitances, respectively; and C_T is the total interconnect capacitance. F is the width of the metal line, and line space = line width.

Another problem is related to the parasitic capacitance of metal interconnects. In spite of the lower metal-to-substrate capacitance (C_o) when compared with bulk Si, SOS has higher metal-to-metal capacitances (C_1 and C_2 in Fig. 5.15) because the high dielectric constant of sapphire ($\varepsilon_{sapphire}$ = 11, ε_{oxide} = 3.9) provides a stronger capacitive coupling between metal lines. This inter-metal capacitance becomes a dominant capacitance as the metal lines and spacings are scaled down for VLSI interconnects. Consequently, as shown in Fig. 5.15, the total interconnect capacitance in SOS is comparable or higher than that of bulk for one-micron CMOS circuits.[19] Additionally, because the interconnect capacitance is dominant in total parasitic capacitance for VLSI, one-micron SOS circuits do not have a speed advantage over their bulk counterparts. Table 5.2 shows a checklist of various considerations for VLSI circuits in favor of CMOS/SOS, bulk CMOS or basically no difference.

TABLE 5.2 Comparison of SOS and Bulk Si Technologies in Various Considerations

Considerations	Favoring SOS	Favoring Bulk	No Difference
Density	x		
Isolation and latch-up	x		
Radiation hardness	x		
Capacitance to Sub.	x		
Soft error	x		
Leakage		x	
Noise margins		x	
Mobility		x	
Substrate defects & floating sub.		x	
Materials cost/availability		x	
Inter-wire capacitance		x	
Total parasitic capacitance			x
VLSI speed			x

5.9 SOI TECHNOLOGY

Insulating substrates other than sapphire have also been used for fabricating CMOS ICs, commonly referred to as silicon-on-insulator (SOI) technology. In this technology, a crystalline Si film is grown on a layer of oxide or nitride formed on Si wafers. The idea is based on inexpensive and readily available substrate. Furthermore, the use of stacked, thin films opens the potential of three-dimensional ICs. Many SOI techniques such as dielectric isolation with substrate removal,[20] Full Isolation by Porous Oxidized Silicon (FIPOS),[21] Silicon IMplanted with OXygen (SIMOX),[22] Epitaxial Lateral Overgrowth (ELO),[23-25] and polysilicon recrystallization[26] have been proposed. Those that are important for CMOS ICs will be discussed in this chapter.

5.9.1 Epitaxy on Insulator

In the SIMOX process shown in Fig. 5.16, high-dose, low energy oxygen or nitrogen atoms are implanted into a Si substrate. At a subsequent high-temperature anneal, these atoms react with Si forming an SiO_2 or Si_3N_4 film

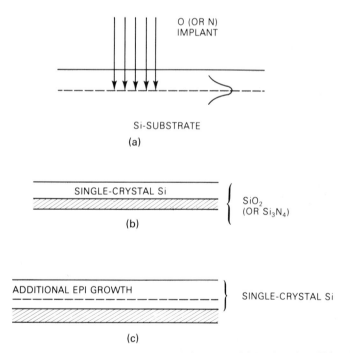

Figure 5.16   Si Implanted OXygen (SIMOX) process. (a) Implant O or N into Si substrate; (b) anneal to SiO_2 or Si_3N_4 under a thin layer of Si; and (c) grow additional epi Si layer.

buried in, but near, the surface of the Si substrate. The thin single-crystal Si layer above the insulator is then used as a seed for growing a thicker Si epi-layer. Good electrical characteristics have been obtained for CMOS devices and circuits built on implanted-buried nitride[27] and oxide[28] wafers (Fig. 5.17). Leakage current is not excessive but still higher than bulk CMOS, especially at high drain biases. The throughput of implanting oxygen or nitrogen to achieve atomic concentrations at 10^{18} or 10^{17} cm^{-3} has been a practical concern. The application of this technology for VLSI was recently made possible through the demonstration of a 4 K bit CMOS SRAM.[29]   In this work, 18 wafers were implanted in 6.5 hours through the use of a high current implanter.

Figure 5.18 shows the process steps for lateral epi growth over oxide.[24] A Si epitaxial layer is grown around the SiO_2 island and laterally expanded over the island. A conventional epi reactor can be used for this process, thus no new equipment is needed. pMOSFETs made with this technique

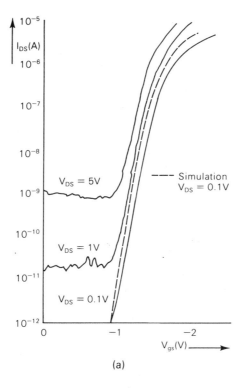

(a)

Figure 5.17 Subthreshold characteristics for MOSFETs made by SIMOX processes. (a) A pMOSFET on buried nitride (Ref. 27, © 1983, IEEE); and (b) CMOS-FETs on buried oxide (Ref. 28, © 1985, IEEE), channel width is 10 μm and V_{DS} = 0.1 V.

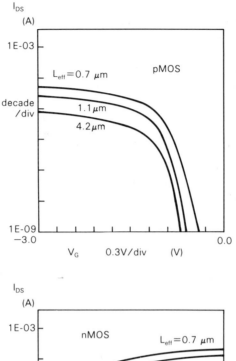

Figure 5.17 *Continued.*

have shown characteristics similar to those in bulk devices.[24,25] However, minority lifetimes for the ELO materials were found to be an order of magnitude lower than those of bulk control wafers. The process is still in an early stage.

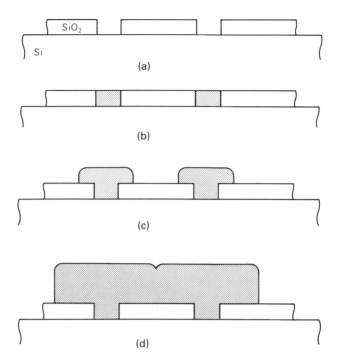

Figure 5.18   Process for Epitaxial Lateral Overgrowth (ELO). (a) Define a seed pattern by etching window in an oxide layer; (b) vertical epi growth from the seed windows; (c) lateral epi overgrowth; and (d) epi growth completed (Ref. 24, © 1983, IEEE).

5.9.2 Recrystallization of Polysilicon

Another SOI technique that has received considerable attention is the recrystallization of polysilicon film. This technique involves recrystallizing, or at least enlarging the grains, of a polysilicon film by scanning a heat source over the film (Fig. 5.19).[30] Heat sources used for this purpose include lasers, electron beams, graphite strip heaters, and incoherent light sources.

CW laser scanning has been effective in forming larger poly grains. The laser scans a small energy spot at a very short (msec) radiation time. The advantages are rapid heat dissipation and self-limiting energy absorption. Thus, disturbance of devices beneath the SOI layer is minimal. The disadvantages are low throughput and difficulty in producing large-area, single-crystal films. It normally needs a capping layer to control the amount of energy absorption and minimize surface-tension effects.

Electron beam scanning is less sensitive to the capping material and its thickness. It also offers higher power density and can emulate a line source

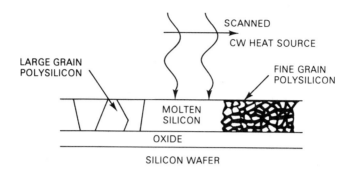

Figure 5.19   Schematic of a scanning heat source for large-grain polysilicon growth.

by scanning the beam rapidly (MHz rates) in the x direction and moving slowly (10 cm/sec) in the y direction.   Grains as large as 50×120 μm² have been achieved with this technique.[31]   Higher throughput is expected, but commercial equipment is not readily available.   Charging of the insulating substrate and e-beam radiation damage are other concerns.   As shown in Table 5.3, CMOSFETs fabricated in laser or e-beam recrystallized films are superior compared to those made in as-deposited poly films and their mobility values approach those of bulk Si.[30]

Black body radiation from a strip heater has produced large-area single crystals in which CMOS devices comparable to bulk devices have been demonstrated.[32]   A sample is first heated to a high background temperature, then a moving line source produced by the strip heater melts poly-Si films and provides lateral epitaxial growth (Fig. 5.20) at a high throughput.   The corresponding high substrate temperature and long radiation time (seconds) however tend to destroy the devices in the substrate.   Contamination from graphite is another consideration.   Incoherent light sources such as arc lamps are probably cleaner than graphite heaters, but only offer limited power.

TABLE 5.3   Characteristics of MOSFETs Fabricated in Single-Crystal Si and in Poly Si (Ref. 30, © 1982, IEEE).

	n-Channel		*p*-Channel	
	Mobility (cm²/V-sec)	V_t (V)	Mobility (cm²/V-sec)	V_t (V)
Single-crystal silicon	670	1	180	-1.5
Recrystallized polysilicon	>300	1–2	>120	-2
Fine-grain polysilicon	~10–20	10–20	~20	-10 to -20

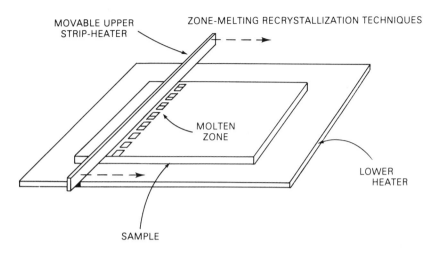

MOVABLE UPPER STRIP-HEATER

ZONE-MELTING RECRYSTALLIZATION TECHNIQUES

MOLTEN ZONE

LOWER HEATER

SAMPLE

Figure 5.20   Schematic of zone-melting recrystalization technique.

5.9.3 CMOS/SOI for 3-D Integration

SOI technology opens up the possibility of fabricating devices in more than one layer, leading to three-dimensional (3-D) integration. In principle, this integration will overcome the shortcoming of the poor packing density associated with CMOS and take the advantage of its low power characteristic, which is absolutely needed in a 3-D IC. Combining CMOS with SOI has the potential of offering a high-density and low-power technology for future VLSI. However, many technological barriers exist in forming an IC with multiple layers.

As previously discussed, laser or *e*-beam recrystallization seems more likely in making 3-D ICs because this technique does not greatly disturb the underlying devices in terms of dopant diffusion and oxide damage. A CMOS inverter with one type of FET in the SOI layer and another type in the substrate has been demonstrated using laser recrystallization. The CMOS inverter has two different configurations.[33] The first employs the idea of flipping the *p*-channel FET on top of the *n*-channel with a two poly gate joint (Fig. 5.21a). This Joint-gate MOSFET (JMOS) has been successfully demonstrated.[34] However, the difficulties in making the top MOSFET with self-aligned S/D and forming a good SOI layer on a thin gate oxide have made this structure hard to scale for VLSI. The second configuration is to simply stack one FET on top of the other with an insulator in between (Fig. 5.21b). Although one more poly layer is used, the difficulties of JMOS formation do not exist. The corresponding process is also more controllable. The stacked structure has been realized with laser[33] and *e*-beam recrystallization.[35]

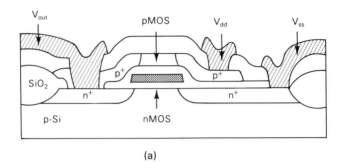

(a)

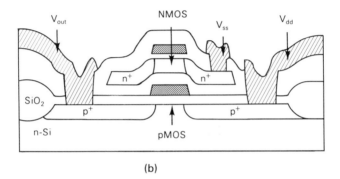

(b)

Figure 5.21 Schematic cross-sections of a 3-D CMOS/SOI inverter: (a) Joint-gate (JMOS) type; and (b) stacked type (Ref. 33, © 1983, IEEE).

To date, a 1.1K gate array[36] and a 256 bit SRAM[37] have been demonstrated for 3-D CMOS/SOI. The 8K bit parallel array multiplier in the gate array exhibits complete operation. The 256-bit SRAM is configurated with the NMOS memory cells in the bottom layer and the CMOS peripheral circuits in the top layer as shown in Fig. 5.22. Complete memory operation including the intralayer and interlayer data transfer has been demonstrated. Both these circuits were fabricated with a laser recrystallized SOI layer. An even larger circuit, a 64K SRAM, was demonstrated[38] using p-channel loads made in a non-crystallized polysilicon SOI layer. The as-deposited polysilicon was hydrogen passivated rather than beam recrystallized for manufacturability. Although the p-channel polysilicon FETs had low mobility, the resultant CMOS configuration provided lower static power than that of a conventional NMOS RAM with poly load resistors. On the other hand, SOI advantages such as latchup free, high density, high alpha-particle immunity also existed in this structure.

Despite the fact that sizable CMOS circuits have been demonstrated in two-layer SOI, 3-D VLSI using CMOS/SOI technology has a long way to go

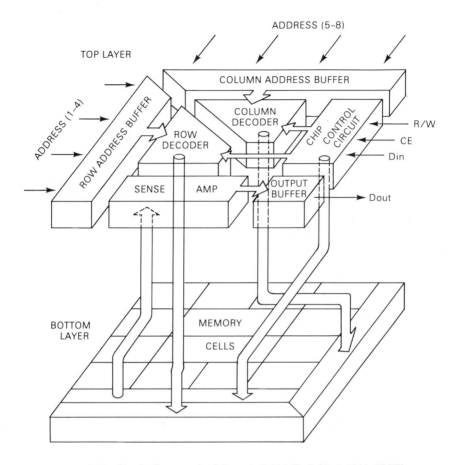

Figure 5.22   Circuit diagram of a 3-D static RAM (Ref. 37, © 1986, IEEE).

for production.   A reliable and high-throughput recrystallization process for SOI layers is needed.   Intra- and inter-layer interconnection with planarization must be developed.   High yield, fast turn-around and redundance circuits are other concerns for VLSI fabrication.

5.10 BIPOLAR/CMOS INTEGRATION—BiCMOS

As described in Chapter 1, CMOS offers low power which is attractive for VLSI, especially in digital applications.   However, the speed of CMOS, although comparable with NMOS, is slower than what bipolar can provide. It is particularly true when heavy-loading or long interconnects need to be driven.   A bipolar transistor not only can deliver a large current, it also has well-controlled turn-on voltage, good noise margin and small logic swing for

ECL (Emitter-Coupled Logic). ECL also provides a high speed sense amplifier for static RAMs. All these advantages make it suitable for high speed
circuits and analog applications. Its major disadvantages are: high power,
poor density and limited circuit options.

BiCMOS technology is the integration of bipolar and CMOS devices
on a single chip. The intent is to offer high density, low power CMOS arrays
with high speed bipolar drivers. It can also provide analog and digital system
integration on the same chip. With BiCMOS, the advantage of high density
and low power in CMOS can be combined with the speed advantage offered
by bipolar. Other improvements include ECL and TTL (Transistor Transistor Logic) interface, high speed I/Os, less sensitivity to fan-out and output
load, reduced clock skew and improved internal gate delay. Because bipolar
devices offer more current driving ability, for high speed applications, Bi-
CMOS does not have to be down-scaled as much as CMOS. One-to-two
micron BiCMOS can offer circuit speed as high as submicron CMOS. Consequently, the 5 V power supply can be maintained and submicron process
needs are not as high. The major drawbacks to BiCMOS are higher cost
and longer fabrication time.

5.10.1 BiCMOS Operation

Figure 5.23 shows a BiCMOS inverter. When the input (V_{in}) is low,
both N0 and N1 are off but the p-channel device is on and the base of the
upper npn bipolar rises to V_{BE} and above, turning the bipolar on and pulling
the V_{out} up to $V_{DD} - V_{BE}$. When the input (V_{in}) is switched to high, Q1 is
cut off, and so is N2; but N1 is turned on and the base of the lower npn
bipolar is charged (V_{out}) to V_{BE}, turning the bipolar on and pulling the output
voltage (V_o) down to V_{BE}.

Because the BiCMOS output does not swing the full rail (V_{DD} to ground),
the superior noise margin and zero static power dissipation inherent in CMOS
are not fully maintained. As the output is pulled down, V_{out} goes to V_{BE};
although V_{out} eventually can be discharged below V_{BE}, it would take an
unreasonably long time. For practical cases, the BiCMOS output does not
go to ground, noise margin is therefore degraded and when it drives to another
CMOS input gate, power dissipation for the CMOS gate is slightly increased.
The amount of power dissipation depends on the threshold of the n-channel
FET in the CMOS input gate. If 1 μA is used to define the threshold voltage
(V_{Tn}), the static current dissipated at the n-channel FET of the CMOS gate
can be approximated as

$$I_D = 10^{-6} - (V_{Tn} - V_{out})/S_t \text{ in Amp} \qquad (5.1)$$

where V_{Tn} is the n-channel FET threshold voltage and S_t is the FET subthreshold factor in mV per decade as described in Chapter 2 (Eq. 2.36). The
static power dissipation of the CMOS gate, which is proportional to I_D, can

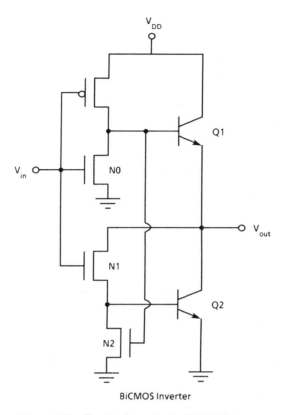

BiCMOS Inverter

Figure 5.23   Circuit schematic of a BiCMOS inverter.

be reduced by increasing the threshold voltage and decreasing subthreshold swing.   However, too high a threshold voltage results in lower current driving ability, hence poorer circuit speed.   As an example, if $V_{Tn} = 0.7$ V, $S_t = 100$ mV/decade, then $I_D \cong 1$ μA because $V_{BE} \cong 0.7$ V.   If V_{Tn} is raised to 1 V, then $I_D \cong 1$ nA, an improvement of three orders of magnitude in power dissipation.

Figure 5.24 shows delay time for the BiCMOS inverter plotted versus load capacitances (C_L).[39]   Notice that the speed improvement is greater as the load capacitance becomes larger.   This improvement is due to the fact that a bipolar device normally delivers much more current than a MOS device.   Therefore, compared to CMOS, bipolar or BiCMOS is less sensitive to the output load capacitance.   In first order approximation, the slope of the line is roughly

$$S_L = t_{pd}/C_L = Q/(I_L C_L) = V_o/I_L$$

$$= V_o/I_D \text{ for the CMOS case} \qquad (5.2)$$

$$= V_o/I_C = V_o/(\beta I_B) \text{ for the BiCMOS case}$$

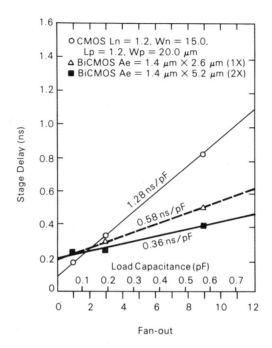

Figure 5.24 Propagation delay versus load capacitance (Ref. 39).

where Q is the charge stored at the load capacitor, V_o is the output voltage, and I_L is the output current, which is the drain current (I_D) for the CMOS case, and is the collector current (I_C) in the BiCMOS case. Although I_D and I_C are geometry-dependent, i.e., $I_D \propto W/L$ and $I_C \propto I_E \propto A_E$, the bipolar current I_C in general is much larger than the drain current I_D of a FET of similar size. So, for gates that need to drive a heavy load (≥ 0.5 pF), BiCMOS is faster for a factor of two or more. For a very small load (<0.1 pF in this example), the BiCMOS circuit is actually slower because it takes time to charge additional bipolar junction capacitances. In integrated circuits, a large fan-out normally corresponds to a high load capacitance. Compared to bipolar for the same fan-out, CMOS has a larger load capacitance to drive due to the higher input capacitance. As shown in Fig. 5.24, delays for CMOS inverters increase as 1.28 ns/pF and this loading effect decreases to 0.58 and 0.36 ns/pF for BiCMOS inverters with two different emitter areas. Notice for fan-out greater than 2, the BiCMOS inverters run faster than the CMOS inverter.

Slightly more complex BiCMOS logic such as NAND and NOR gates are shown in Fig. 5.25. Again, only two bipolar devices are used in spite of the increase in FET counts. With respect to the total device count and extra

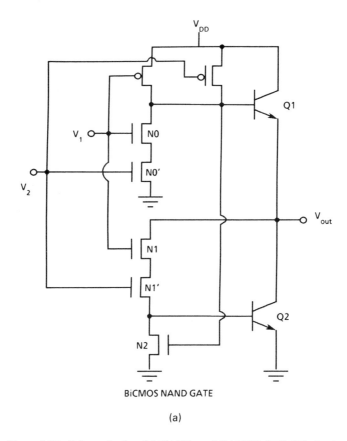

BiCMOS NAND GATE

(a)

Figure 5.25   Schematics for: (a) NAND; and (b) NOR BiCMOS circuits.

layout area, the penalty of adding bipolar drivers on a CMOS logic circuit is reduced as the CMOS circuit component becomes larger.

5.10.2 Bipolar Device Characteristics

In this section, one-dimensional bipolar device operation is discussed. Important device characteristics and relevant process parameters are also described to provide a background for the discussion of BiCMOS technology.

Shown in Fig. 5.26 are the cross-sectional view of an *npn* bipolar transistor, its corresponding one-dimensional representation and the impurity doping distribution. It is a three terminal device with a heavily doped *n*-type emitter that emits current, a lightly doped *n*-type collector that collects current and a *p*-type base region. As shown in Fig. 5.26(b), it is like two *p-n* junctions connected back-to-back, but the *p* region is very narrow. During normal operation, the emitter-base *pn* junction is forward biased and the

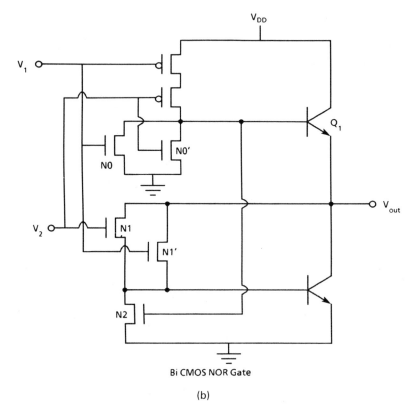

Bi CMOS NOR Gate

(b)

Figure 5.25   *Continued.*

collector-base junction is reverse biased.   A transistor biased under this condition operates in the so-called active region.   Under forward bias, the emitter
current (I_E) is caused by the diffusion of electrons from the n^+ emitter to the
p-type base (I_{NE}) and the diffusion of holes from the base to the emitter (I_{PE}).
The current components governed by diffusion are

$$I_{NE} = A_E q D_{nB} \, dn'_B/dx = A_E q D_{nB} \, n'_B(0)/W_B \tag{5.3a}$$

$$I_{PE} = A_E q D_{pE} \, dp'_E/dx = A_E q D_{pE} \, p'_E(0)/W_E \tag{5.3b}$$

$$I_E = I_{NE} + I_{PE} \tag{5.3c}$$

assuming W_B and W_E are very small so that linear approximation can be
applied. A_E is the emitter area, D_n and D_p are the diffusivities for electrons
and holes, n'_B and p'_E are the excess minority carrier concentrations in the
base and emitter regions due to injection under forward bias at the emitter-

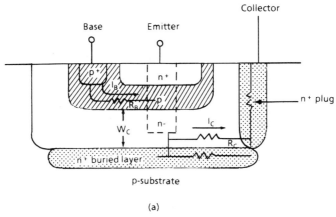

(a)

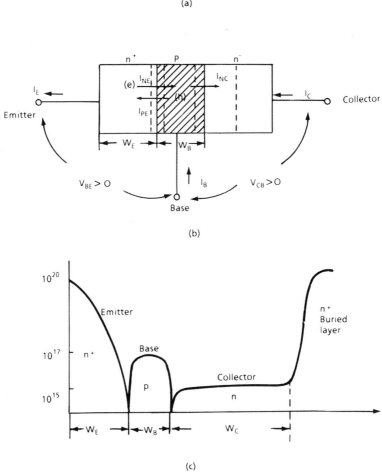

(b)

(c)

Figure 5.26 An *npn* bipolar transistor: (a) cross-section of an *npn* transistor; (b) One-dimensional representation; and (c) Impurity doping distribution in one dimension.

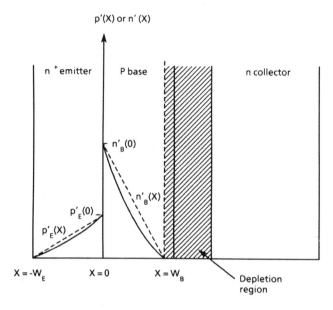

Figure 5.27   Excessive minority carrier distribution in a one-dimensional *npn* transistor.

base junction. They are calculated as

$$n_B'(x) = n_B(x) - n_{B0} = n_{B0}[\exp(qV_{BE}/kT) - 1] \quad \text{at } x = 0 \qquad (5.4a)$$

$$p_E'(x) = p_E(x) - p_{E0} = p_{E0}[\exp(qV_{BE}/kT) - 1] \quad \text{at } x = 0 \qquad (5.4b)$$

where n_B and p_E are the minority carrier concentrations, n_{B0} and p_{E0} are the minority concentrations at equilibrium, and V_{BE} is the forward bias emitter-base voltage. Figure 5.27 shows the distribution of minority concentrations in the base and emitter regions. The slopes of the curves correspond to current values. The excess electron concentration reaches zero at the base-collector junction because the junction is reverse biased and the electric field in the depletion region across the junction sweeps carriers away. Notice that $n_B'(0)$ is much larger than $p_E'(0)$ because $pn = n_i^2$ and the majority carrier concentration (i.e., the doping concentration) in the base is much less than that in the emitter. As a result, $I_{PE} \ll I_{NE}$. The ratio of the current injected into the base to the total emitter current is defined as emitter efficiency (γ_e) and is expressed as

$$\gamma_e = I_{NE}/(I_{NE} + I_{PE}) \cong 1 - I_{PE}/I_{NE}$$
$$= 1 - [D_p p_E'(0)W_B]/[W_E D_n n_B'(0)] \qquad (5.5)$$

Using Einstein relation ($D = \mu kT/q$), the above equation can be written as

$$\gamma_e = 1 - (\mu_p p_B W_B)/(\mu_n n_E W_E) \cong 1 - R_{SE}/R_{SB} \qquad (5.6)$$

where p_B and n_E are majority carrier densities in the base and the emitter. $p_B W_B$ is the total majority charge in the base and is called the Gummel number. The higher the Gummel number, the lower the emitter efficiency. R_{SE} and R_{SB} are the sheet resistances for both the emitter and base, respectively. Because typical values of R_{SE} and R_{SB} are tens and thousands respectively, γ_e is close to but less than unity.

Not all injected electrons are collected at the collector. Some of them are recombined in the base region and the portion which reaches the collector-base junction is defined as the base transport factor (T):

$$T = \frac{I_{NC}}{I_{NE}} = \frac{(dn_B/dx)_{x=W_B}}{(dn_B/dx)_{x=0}} \tag{5.7}$$

Solving the diffusion equation for n_B with boundary conditions, the above equation can also be shown as[40]

$$T = (\mathrm{Cosh} W_B/L_{nB})^{-1}$$
$$= 1 - W_B^2/(2L_{nB}^2) \quad \text{since } W_B \ll L_{nB} \tag{5.8}$$

where L_{nB} is the diffusion length of minority carriers in the base. $L_{nB} = (D_n \tau_n)^{1/2}$ where τ_n is the minority lifetime.

Because the collector-base junction is reverse biased, the total collector current (I_C) is approximately equal to I_{NC} assuming the junction leakage current is negligible. The common base current gain (α) can then be written as

$$\alpha = \frac{I_C}{I_E} = \frac{I_{NC}}{I_{NE}} \frac{I_{NE}}{I_E} = T\gamma_e = [1 - W_B^2/(2L_{nB}^2)](1 - R_{SE}/R_{SB}) \tag{5.9}$$

Again, α is typically slightly less than one.

Based on Eqs. 5.3a and 5.4a, the current driving ability is

$$I_C = \alpha I_E = \alpha I_{NE}/\gamma_e = \frac{T A_E q D_n n_i^2 [\exp(qV_{BE}/kT) - 1]}{p_B W_B} \tag{5.10}$$

The denominator is the Gummel number. The numerator indicates that the current of a bipolar transistor is exponentially dependent on the voltage across the emitter-base junction. In reality, because of the IR drop from the base contact to the emitter-base junction (Figure 5.26a), the V_{BE} term in the above equation should be replaced by $V_{BE} - I_B R_B$.

The I_C versus V_{BE} curve is known as the Gummel plot. Figure 5.28 shows experimental data for I_C and I_B versus V_{BE} plotted linearly and semi-logarithmically. Notice that the curve in Figure 5.28(b) is steep and a turn-on voltage at about 0.7 V can be defined. A larger emitter area, a narrow base and/or a lower base doping concentration can increase bipolar current driving ability. However, a narrow base or a low base doping concentration can cause significant base width modulation, an undesirable effect similar to

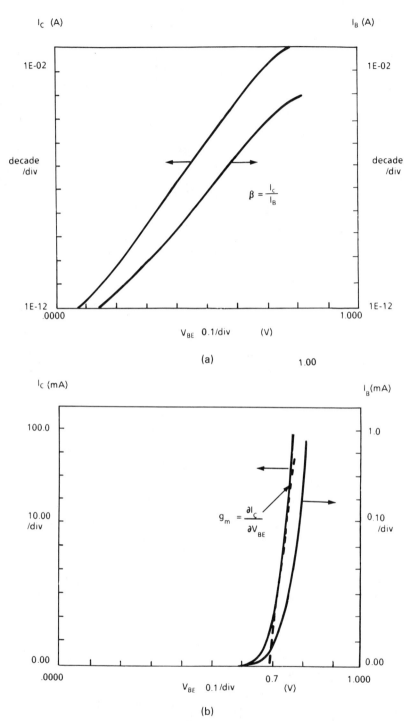

Figure 5.28 Collector and base current (I_C and I_B) versus base-emitter voltage (V_{BE}) for an *npn* transistor: (a) Gummel plot in semi-logarithmic; and (b) Gummel plot in linear.

the short-channel effect in MOSFETs. This base width modulation, known as Early effect, leads to increased output conductance for the bipolar transistor.

The base doping concentration should be much higher than the collector doping concentration so that most of the depletion width extends to the collector region to minimize Early effect and to maintain sufficient collector-base breakdown voltage (BV_{CBO}) and collector-emitter breakdown voltage (BV_{CEO}). However, a lightly doped collector gives a high collector resistance (R_C), resulting in a reduced current at a given bias condition. A common way to accommodate both high breakdown and low collector resistance is to use a heavily doped buried layer (Fig. 5.26a) under the lightly doped collector region so that R_C can be greatly reduced.

The common emitter current gain (β) is defined as I_C/I_B and equals $\alpha/(1 - \alpha)$. Depending on α, β values range from tens to hundreds. The higher the α, the higher the β. From Eq. 5.9, it is clear that narrow base width, low emitter sheet resistance and low Gummel number in the base can all increase the common emitter current gain. Again, the Gummel number cannot be too low due to the limitation in output conductance and breakdown voltages.

While current gains are important DC parameters, transit time (τ_t) is important for high frequency operation. τ_t is the time needed for the carriers to travel across the base region. It can be approximated by

$$\tau_t = W_B/\langle v \rangle \tag{5.11}$$

where v is carrier velocity and $\langle\ \rangle$ means average value. Because

$$I_C = A_E q \langle n'_B \rangle \langle v \rangle = A_E q \langle v \rangle n'_B(0)/2 \tag{5.12}$$

and

$$I_C \cong I_{NE} = q D_n A_E n'_B(0)/W_B \tag{5.13}$$

equating Eq. (5.12) and (5.13),

$$\langle v \rangle = 2D_n/W_B \tag{5.14}$$

and

$$\tau_t = W_B^2/2D_n \tag{5.15}$$

can be obtained.

The above derivation, based on the concepts of average velocity and average minority carrier concentration, is at best an estimation. τ_t is also the time required to remove all minority carriers from the base when the emitter-base voltage (V_{BE}) switches to zero. It is interesting to note that $\tau_n/\tau_t \approx \beta$. The reciprocal of the transit time corresponds to the frequency limitation for the operation of a bipolar transistor. In reality, even before the operating frequency reaches $1/\tau_t$, current gain decreases at high frequencies due to additional RC time constants associated with parasitics. The cutoff frequency f_T, defined as the frequency at which the common emitter current gain is unity, is an important parameter for high frequency operations. f_T can be expressed as[41]

$$f_T = 1/(2\pi\tau_{total}) \tag{5.16}$$

and

$$\tau_{total} = \tau_t + C_{jE}/g_m + C_{jC}/g_m \tag{5.17}$$

where C_{jE} and C_{jC} are the junction capacitances at emitter and collector junctions and g_m is the transconductance defined as $\partial I_C/\partial V_{BE}$. To obtain good high frequency response, the parasitic capacitances and resistances must be reduced. In addition, the intrinsic base transit time must be decreased. Along with the use of the buried layer, a thinner collector layer (small W) and/or a heavily doped collector plug under the collector contact can further decrease R_C (Fig. 5.26a).

The previous discussion assumes that the carrier concentration injected to the base is much less than the majority carrier concentration (p_B) which equals the base doping concentration (N_{AB}). This assumption is often invalid at high current levels in which $n_B >> N_{AB}$. In such cases, the majority carrier concentration must increase to maintain charge neutrality in the base. This condition is called conductivity modulation. As a consequence, the effective base doping concentration becomes higher, resulting in a smaller emitter efficiency and a reduced current gain. High level injection also occurs at the collector and is known as Kirk effect.[42] For *npn* devices, it was assumed that minority carriers (electrons) that reached the depletion edge of the reversely biased collector-base junction were swept across so that electron density (n) in the depletion region was negligible. This assumption is not valid at a high current density J ($J = qnv$ where v is electron velocity). Because v is limited by the saturation velocity due to scattering, a finite amount of n exists when J is large. This electron density J/v adds to the negatively charged ionized acceptors in the depletion region at the base side so that the total charge density increases.

At the collector side, where the depletion region was composed of the positively charged ionized donors, the injected electrons decreased the charge density and even changed the net charge from positive to negative. In fact, the concept of the depletion region is not reflected accurately here due to the presence of a large number of carriers. It should be described as a space charge region. This modulation of charge density, for a constant collector-base voltage, must be accompanied by a narrowing of the depletion region at the base and a widening of the depletion region at the collector side. Consequently, the undepleted (or neutral) base widens, resulting in a reduction of the current gain at high currents and a poorer frequency response.

5.10.3 BiCMOS Process Technology

BiCMOS technology can be developed in two basic ways: adding bipolar on CMOS or adding CMOS on bipolar.   Because the discussion, thus far, has been concentrated on CMOS technology, the former method will be used to build BiCMOS.   This method has been used for several recent BiCMOS developments.[43-46]   It often compromises bipolar performance to be compatible with existing CMOS and minimize added process steps.   In a BiCMOS circuit, vertical *npn* transistors are needed because they have higher current gain compared to a *pnp* transistor or a lateral device.   Furthermore, each bipolar transistor should be isolated so that it is not limited to emitter-follower type logic only.   *N*-wells are used in CMOS as the isolated collectors for the vertical *npn* transistors.   Thus, CMOS with *n*-well[44] or twin-tub on *p*-substrate[46] is commonly extended to BiCMOS.

Because buried-layer is normally used for *npn* bipolars to reduce collector resistance, a buried-layer mask is defined and followed by a shallow arsenic implant (Fig. 5.29a).   A *p*-type epi layer is then grown and *n*-wells are formed above the n^+ buried layer (Fig. 5.29b).   After the standard CMOS LOCOS field isolation, another mask and boron implant are then needed to form the *npn* base (Fig. 5.29c).   Subsequent to standard gate oxidation, polysilicon deposition and gate definition, the n^+ mask is used to implant arsenic for the *n*MOS S/D, the *npn* emitter and n^+ contacts for the collector and *n*-wells (Fig. 5.29d).   Similarly, p^+ mask is used to implant boron for the *p*MOS S/D and base contact.   Standard contact and metal masks and associated process steps finish the entire BiCMOS process (Fig. 5.29e).   Notice that Masks 1 and 4 are the additional masks needed for making bipolar on a CMOS chip.   With buried-layer, it is necessary to grow an epi layer, which is often already included in CMOS for latch-up prevention.   Boron base implant is yet another extra process step.   The entire process consists of 9 masks: buried layer, *n*-well, active area, *npn* base, poly, n^+, p^+, contact and metal.   This sequence described the basic BiCMOS process in which MOSFETs are not scaled to minimize short-channel effects and *npn* bipolars are not high performance devices.

As discussed in Secs. 5.2 and 5.3, for micron or submicron CMOS, additional masks and implants are necessary for FET threshold adjustment and punchthrough protection, as well as for the double-level metal.   The boron implant designed for *n*-channel punchthrough protection can also be used to form the *npn* base; however, this compromise does not, in general, produce a base well suited for the *npn* bipolar.   To incorporate *n*-well and n^+ buried layer, the BiCMOS substrate must be a lightly-doped *p*-type.   For latchup immunity, p^+ buried-layer can also be included to reduce the substrate resistance (shown as a dashed curve in Fig. 5.24e).   The p^+ layer also creates a p/p^+ hi-low junction which is known to reduce α-particle induced soft error

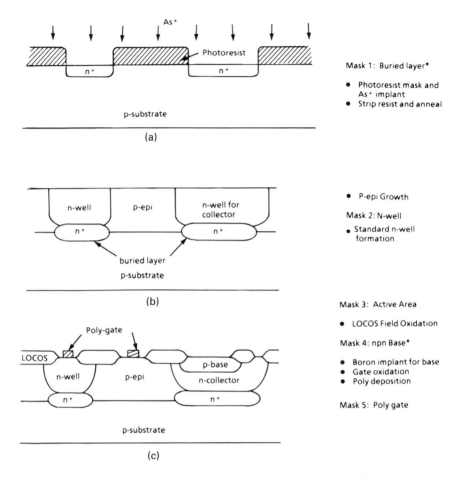

Figure 5.29 An n-well BiCMOS process: (a) buried layer formation; (b) epi growth and n-well formation; (c) active area & base formation and poly gate definition; (d) n^+; and (e) p^+ and contact & metal.

as described in Chapter 4. This addition requires one more mask, making the BiCMOS process three masks more than the twelve masks commonly used for today's CMOS processes. Another significant performance improvement is the addition of a polysilicon emitter[45,46] which is known to increase emitter efficiency, hence current gain. Because polysilicon technology has matured in CMOS, it would not be difficult to add to the bipolar devices. In fact, for SRAMs using poly-load resistors, the same polysilicon layer can be used for the poly emitters. Trench isolation is very attractive for this hybrid technology because several different types of devices need to be isolated with tight isolation space. Figure 5.30 shows an example of the

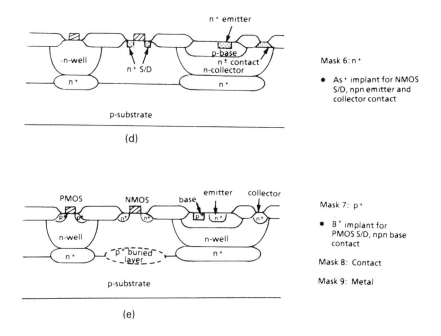

Figure 5.29   *Continued.*

advanced BiCMOS device structure with trench isolation, n^+ and p^+ buried layer, and a polysilicon emitter.

5.10.4 BiCMOS Applications

Static random access memory.   BiCMOS is penetrating the CMOS market in high speed applications.   Because SRAMs are always performance-sensitive, BiCMOS is attractive for SRAMs requiring higher speed, higher density and lower power than what can currently be achieved by either CMOS or ECL.   While CMOS cannot compete on speed, pure ECL dissipates too much power and is very costly.   With BiCMOS SRAM, sub-10 ns access time and <1 W power dissipation are feasible.   These characteristics offer BiCMOS an opportunity in high performance workstation and superminicomputer applications.   One example is a second-level cache to the memory architecture that serves as a backup for higher system speeds by supporting the first-level (CPU) cache on misses and reducing the need to access the main memory, which normally consists of DRAMs with much longer (100–150 ns) access times.   This second-level cache is best implemented by large (1 Mwords) SRAMs with cache cycle times set at 25 ns.   256K BiCMOS

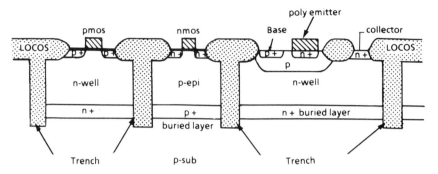

Figure 5.30 Schematic cross-section of an advanced BiCMOS device structure with trench isolation and polysilicon emitter.

SRAMs with 10–15 ns access time and low power dissipation are desirable for this unique application.

Most present BiCMOS SRAMs use conventional 6-T CMOS cells or 4-T resistor-load nMOS cells with bipolar transistors only for sense amplifiers and peripheral circuits.The BIT line and the $\overline{\text{BIT}}$ line from a CMOS RAM cell are connected to a bipolar differential pair of the sense amplifier for high speed sensing. The bipolar differential amplifier is similar to the CMOS amplifier discussed in Chapter 4. The two nMOS input FETs are now re-

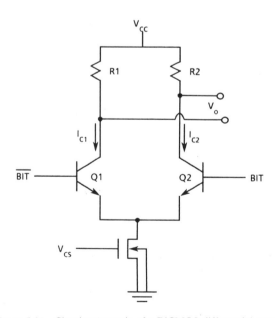

Figure 5.31 Circuit schematic of a BiCMOS differential amplifier.

placed by the two *npn* bipolar devices shown in Fig. 5.31.   The load resistors can be implemented by *p*MOSFETs.   Compared to the CMOS differential pair in which the FET drain current was linearly proportional to the gate voltage, the bipolar collector current (I_{c1} or I_{c2}) is exponentially dependent on the base-emitter voltage shown in Eq. 5.10.   As a result, the bipolar differential amplifier has much higher gain for a very small differential input signal, thereby leading to a very fast sense amplifier.

BiCMOS SRAMs containing CMOS or *n*MOS cells have been made at 256 kbit complexity.[47-49]   Table 5.4 summarizes their characteristics and compares them to high speed ECL RAMs.[50-52]   BiCMOS access times are 10 ± 2 ns, not much slower than ECL speed, the power dissipation is much less, and the cell sizes are much smaller.   The Hitachi ECL uses a Ta_2O_5 capacitor and the Fujitsu ECL uses a *pnp* load cell so that memory cell size can be kept around 500 μm^2.   BiCMOS technologies for the 256K SRAMs have current gain (β) ranging from 80–100 and f_T's at 8 ± 1 GHz.   For comparison, 1 Mbit CMOS SRAMs made by ~0.8 μm technology have access times of 15–25 ns,[53-56] with cell sizes of ~60 μm^2 for 6-T cells.

BiCMOS SRAM cells have been proposed[57,58] to decrease access time while maintaining low power of CMOS memory arrays.   Using a bipolar transistor to deliver more current can reduce the time to charge BIT line capacitance during READ operation, but at the expense of larger cell area for the bipolar device.   A two-port BiCMOS SRAM cell that combined ECL level word-line voltage swings and emitter-follower bit line coupling with a static CMOS latch was designed and fabricated.[58]

As shown in Fig. 5.32, CMOS cross-coupled latch is formed by M1, M2, M3 and M4.   The READ Word Line (RWL) acts as a power supply for the CMOS latch, and the WRITE Word Line (WWL) controls the pass transistor (M5) just like a CMOS SRAM cell, but with a single-ended WRITE. WRITE Bit Lines (WBLs) for unselected columns have to be biased at a safe level which does not change the state of the cell residing in the selected row. The condition of disturbing the unselected cells is more severe when RWL is low because of the reduced noise margin associated with the small supply on the CMOS latch.

The cell is read by raising RWL.   The emitter follower (Q6) and Q7 form a differential pair for the sense amplifier.   For a low output, the *npn* bipolar device Q6 is off and the bit line current I_{BL} flows through Q7.   The $I_{BL}R$ drop results in a low V_o.   For high output, the increase of RWL is coupled to the base of Q6 through the *p*MOS (M2).   Q6 is then switching I_{BL} away from Q7 and V_o is at logic high.

An experimental 4K × 1 bit two-port SRAM[58] fabricated by a 1.5 μm 5 GHz BiCMOS process offers a READ access time of 3.8 ns, a power dissipation of 550 mW, and the cell area is 650 μm^2, approximately 35% larger than that of a 6-transistor CMOS cell implemented with the same design rules.   The memory can offer independent and simultaneous READ and

TABLE 5.4 256K BiCMOS SRAMs and ECL SRAMs

Source	Architecture	Technology	Design Rule	Cell Size in μm²		Access time (ns)	Power (W)	β	f_t in GHz
National[47]	256 k × 1	BiCMOS	1.0 μm	96	4-T	12	0.72	80	7
TI[48]	256 k × 1	BiCMOS	0.8 μm	117	6-T*	8	0.75	80	7
Hitachi[49]	256 k × 1	BiCMOS	1.0 μm	54.7	4-T	8	0.4	100	9
Hitachi[50]	4 k × 4	ECL	1.0 μm	495	(Ta_2O_5)	3.5	2		8
Fujitsu[51]	64 k × 1	ECL	1.2 μm	524	(*pnp*)	10	1.3	100	8
IBM[52]	512 k × 10	ECL	1.2 μm	760	(SBD)†	1	2.4	80	5

*Using TiN for self-aligned local interconnect

†Using Schottky Barrier Diode

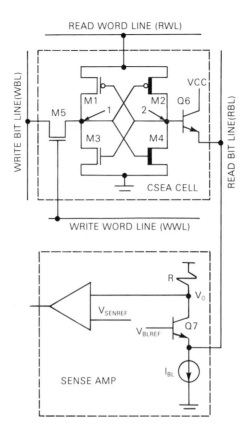

Figure 5.32 Circuit schematic of a two-port BiCMOS SRAM cell (Ref. 58, © 1988, IEEE).

WRITE, making it applicable for the design of video, cache, and other application-specific memories.

Gate array and standard cells.   Combining bipolar and CMOS offers I/O flexibility (ECL and TTL compatible) and new architectures, in addition to high speed and high density in gate array applications.[59] Bipolar provides high driving ability with reduced layout area, making it attractive for high density gate arrays.   Internal CMOS logic with a bipolar buffer results in negligible static power dissipation and low propagation delay under loading. Because high density gate arrays normally use fan-out of 3–5 or even larger, BiCMOS certainly has an advantage in speed as previously described.   For large arrays, BiCMOS has less clock skew because of its high driving ability per unit area; deep clock trees are not needed in general.   Large signal swing in CMOS often results in noise spikes. BiCMOS offers ECL I/Os with small signal swing. Switching noise can then be better controlled. BiCMOS, CMOS RAM cells, ECL buffers, macrocells, and analog functions can all be integrated in application-specific gate arrays.

For a standard cell, BiCMOS also offers a large degree of freedom in cell design and high speed ECL blocks can be developed and integrated with CMOS/BiCMOS logic. Finally, the merge of *npn* bipolar and CMOS transistors can offer high speed and low offset operational amplifiers and comparators. As previously described, a bipolar differential pair offers high gain and more precisely controlled input voltages as compared with its CMOS counterparts. Compared to MOS technology in which threshold voltage is sensitive to manufacturing process and device dimensions, the bipolar emitter-base turn-on voltage (V_{BE}) is much more precisely controlled and well-matched pairs can be more easily obtained. Offset voltages of MOS operational amplifiers are typically an order of magnitude larger than those of bipolar circuits[60] because closely matched MOS input devices are difficult to make and voltage gains are lower. On the other hand, MOS circuits can be made with higher packing density. BiCMOS thus offers the low input offsets and high gains of bipolar circuit components while maintaining high packing density and low power. These components are the basic blocks for high speed and low power A/D and D/A converters. BiCMOS certainly provides more opportunities in the integration of digital and analog functions on a single VLSI chip.

5.10.5 Summary

In general, about three extra masks and associated processes are required for adding bipolar transistors on a CMOS IC. Major gains are: faster speed, higher current driving ability, ECL I/O compatibility, and more flexibility in analog design. Longer turn-around time and higher costs are expected. It is not worth using BiCMOS to replace CMOS for implementing digital systems with low or medium performance. For high performance digital applications or mixed digital/analog systems, additional values offered by BiCMOS justify its extra cost. Moreover, for state-of-the-art CMOS VLSI technology, the mask count is about a dozen and the process steps are already large; another three masks and additional process steps needed for making bipolar devices do not comprise more than 15–20% of the CMOS process steps. In addition, as bipolar and CMOS technology approach submicron dimensions, they share more processing techniques such as epitaxy, polysilicon gate/emitter, trench isolation, and RIE. This technological convergence[61] further justifies the merging of bipolar and CMOS for high performance analog/digital system applications.

REFERENCES

1. K. Lee and A. R. Neureuther, "SIMPL-2 (SIMulated Profiles from the Layout—Version 2)," Technical Digest Symp. on VLSI Technology, Kobe, Japan, p. 64, 1985.

2. L. C. Parrillo, R. S. Payne, R. E. Davis, G. W. Reutlinger, and R. L. Field, "Twin-tub CMOS: A technology for VLSI circuits," in *IEDM Tech. Dig.*, p. 752, 1980; also L. C. Parrillo, L. K. Wang, R. D. Swenumson, R. L. Field, R. C. Melin, and R. A. Levy, "Twin-tub CMOS II: An advanced VLSI technology," in *IEDM Tech. Dig.*, p. 706, 1982.

3. J. Agraz-Guerena, R. Ashton, W. Bertram, R. Melin, R. Sun, and J. T. Clemens, "Twin Tub III—A third generation CMOS technology," in *IEDM Tech. Dig.*, p. 63, 1984.

4. R. D. Rung, C. J. Dell'Oca, and L. G. Walker, "A retrograde *p*-well for high-density CMOS," *IEEE Trans. Electron Devices*, vol. ED-28, p. 1115, 1981.

5. S. R. Combs, "Scalable retrograde *p*-well CMOS technology," in *IEDM Tech. Dig.*, p. 346, 1981.

6. J. Y. Chen, "Quadruple-well CMOS for VLSI technology," *IEEE Trans. on Electron Devices*, vol. ED-31, p. 910, 1984.

7. R. Chwang and K. Yu, "CHMOS: An *n*-well bulk CMOS technology for VLSI," *VLSI Design*, 4th quarter, p. 42, 1981.

8. H. Shichijo, "A re-examination of practical performance limits of scaled *n*-channel and *p*-channel MOS devices for VLSI," *Solid State Electronics*, vol. 26, p. 969, 1983.

9. A. Schmitz and J. Y. Chen, "Design, modelling and fabrication of subhalf-micrometer CMOS transistors," *IEEE Trans. Electron Devices*, vol. EO-33, p. 148, 1986.

10. K. Yu, R. Chwang, M. Bohr, P. Warkentin, S. Stern, and C. N. Berglund, "HMOS-CMOS: A low power high performance technology," *IEEE Journal of Solid State Circuits*, vol. SC-16, p. 454, 1981.

11. R. A. Martin and J. Y. Chen, "Optimized retrograde *n*-well for one micron CMOS technology," *IEEE J. Solid-State Circuits*, vol. 21, p. 286, 1986; also, A. G. Lewis, J. Y. Chen, R. A. Martin, and T. Y. Hunag, "Device isolation in high density CMOS," in the Proc. of the 16th European Solid State Device Res. Conf., Cambridge, U.K., 1986.

12. J. O. Borland, M. Kuo, J. Shibley, B. Roberts, R. Schindler, and T. Dalrymple, "An intrinsic gettering process to improve minority carrier lifetimes in MOS and bipolar silicon epitaxial technology," Semiconductor Processing, ASTM STP 850, C. Gupta, Ed., American Society for Testing and Materials, 1984.

13. L. G. Heller, W. R. Griffin, J. W. Davis, and N. G. Thomas, "Cascade voltage switch logic: A differential CMOS logic family," in *ISSCC*, p. 16, 1984.

14. K. Kokkonen, and R. Pashley, "Modular approach to CMOS technology tailors process to application," *Electronics*, p. 129, 1984.

15. S. Kohyama, J. Matsunaga, and K. Hashimoto, "Directions in CMOS technology," in *IEDM Digest*, p. 151, 1983.

16. Y. Taur, G. H. Hu, R. H. Dennard, L. M. Terman, C. Y. Ting, and K. E. Petrillo, "A self-aligned 1-μm-channel CMOS technology with retrograde *n*-well and thin epitaxy," *J. of Solid-State Circuits*, vol. SC-20, p. 123, 1985.

17. D. C. Mayer, P. K. Vasadev, J. Y. Lee, Y. K. Allen, and R. C. Henderson, "A short channel CMOS/SOS Technology in recrystallized 0.3-μm thick silicon-on-sapphire films," *IEEE Trans. Electron Devices*, vol. EDL-5, p. 156, 1984.

18. D. Leong, "Silicon on sapphire," Silicon on Insulator Technologies for Integrated-Circuit Applications, U.C. Berkeley short course, Oct. 1983.

19. E. Sun, "Performance comparison of MOS transistors and circuits fabricated using bulk and SOS technology," *CICC*, May 1980.

20. K. Bean and W. Runyan, "Dielectric Isolation: comprehensive, current, and future," *J. Electrochem. Soc.*, vol. 124, p. 5C, 1977.

21. Y. Arita and Y. Sunohara, "Formation and properties of porous silicon film," *J. Electrochem. Soc.*, vol. 124, p. 285, 1977.

22. S. Nakashima and K. Okwada, "Electrical characteristics of an upper interface on a buried SiO_2 layer formed by oxygen implantation," *Japan J. Appl. Physics.*, vol. 22, p. 1119, 1983.

23. D. D. Rathman, D. J. Silversmith, and J. A. Burns, "Lateral epitaxial overgrowth of silicon on SiO_2," *J. Electrochem. Soc.*, vol. 129, p. 2303, 1982.

24. L. Jastrzebski, A. C. Ipri, and J. F. Corboy, "Device characterization on monocrystalline silicon grown over SiO_2 by the ELO (Epitaxial Lateral Overgrowth)," *IEEE Electron Device Lett.*, EDL-4, p. 32, 1983.

25. T. I. Kamin and D. R. Bradbury, "Trench-isolated transistors in lateral CVD epitaxial silicon-on-insulator films," *IEEE Electron Device Lett.*, EDL-5, p. 449, 1984.

26. J. Maserjian, "Single-crystal germanium films by microzone melting," *Solid State Electronics*, vol. 6, p. 4477, 1963.

27. G. Zimmer and H. Vogt, "CMOS on buried nitride—a VLSI SOI technology," *IEEE Trans. on Electron Devices*, vol. ED-30, p. 1515, 1983.

28. K. Hashimoto, T.I. Kamins, K. M. Cham, and S. Y. Chiang, "Characteristics of submicrometer CMOS transistors in implanted- buried-oxide SOI films," in *IEDM Dig.*, p. 672, 1985.

29. Y. Omura, S. Nakashima, and K. Izumi, "A 4kb CMOS/SIMOX SRAM," *Dig. of the Sym. on VLSI Technology*, p. 24, 1985.

30. T. I. Kamin, "MOS Transistors in beam-recrystallized polysilicon," *IEDM Digest*, p. 420, 1982.

31. D. B. Rensch and J. Y. Chen, "Recrystallization of Si films on thermal SiO_2—coated Si substrates using a high-speed *e*-beam line source," *IEEE Electron Device Lett.*, vol. EDL-5, p. 38, 1984.

32. M. W. Geis, H. I. Smith, B. Y. Tsaur, S. C. Fan, E. W. Maby, and D. A. Antoniadis, "Zone-melting recrystallization of encapsulated silicon films on SiO_2—

Morphology and crystallography," *Appl. Phys. Lett.*, vol. 40, p. 158, Jan. 1982, and references therein.

33. Motoo Nakano, "3-D SOI/CMOS," in *IEDM Dig.*, p. 792, 1984; also S. Kawamura, N. Sasaki, T. Iwai, M. Nakano, and M. Takegi, "Three-dimensional CMOS IC's fabricated by using beam recrystallization," *IEEE Trans. Electron Device Lett.*, vol. EDL-4, p. 366, 1983.

34. J. F. Gibbons and K. F. Lee, "One-gate-wide CMOS inverter on recrystallized polysilicon," *IEEE Electron Device Lett.*, vol. EDL-1, p. 117, 1980.

35. S. Akiyama, M. Yoneda, S. Ogawa, N. Yoshii, and Y. Terui, "Fabrication technologies for multilayer CMOS device," *Dig. Sym. on VLSI Technology*, p. 28, 1985.

36. K. Sakashita, T. Nishimura, S. Kusunoki, Y. Kuramitsu, and Y. Akasaka, "1.1 K-gate CMOS/SOI gate array by laser recrystallization technique," *Dig. of Sym. on VLSI Technology*, p. 32, 1985.

37. Y. Inoue, K. Sugahara, S. Kusunoki, M. Nakaya, T. Nishimura, Y. Horiba, Y. Akasaka, and H. Nakata, "A three dimensional static RAM," *IEEE Electron Devices Lett.*, EDL-7, p. 327, 1986.

38. S. D. S. Malhi, R. Karnaugh, A. H. Shah, L. Hite, P. K. Chatterjee, H. E. Davis, S. S. Mahant-shetti, C. D. Gosmeyer, R. S. Sundarean, C. E. Chen, H. W. Lam, R. A. Haken, R. F. Pinizzotto, and R. K. Hester, "A VLSI suitable 2-μm stacked CMOS process," presented at 1984 Device Research Conf., abstract published in *IEEE Trans. Electron Devices*, vol. ED-31, p. 1981, 1984.

39. M. P. Brassington, M. El-Diwany, P. Tuntasood and R. R. Razouk, "An advanced submicron BiCMOS technology for VLSI applications," in the *Digest of Sym. on VLSI Technology*, p. 89, 1988.

40. A. S. Grove, *Physics and Technology of Semiconductor Devices*, John Wiley and Sons, Inc., N.Y. 1967.

41. R. S. Muller and T.T. Kamins, *Device Electronics for Integrated Circuits*, ed. by John Wiley & Sons, N.Y. 1977.

42. C. T. Kirk, "A theory of transistor cutoff frequency (f_T) fall off at high current densities," *IEEE Trans. Electron Devices*, ED-9, p. 164, 1962.

43. H. Higuchi, G. Kitsukawa, T. Ikeda, Y. Nishio, N. Sasaki, and K. Ogiue, "Performance and structures of scaled-down bipolar devices merged with CMOS-FETs," *IEDM Dig.*, p. 694, 1984.

44. J. Miyamoto, S. Saitoh, J. Momose, H. Shibata, K. Kanzaki, and S. Kohyama, "A 1.0 μm N-well CMOS/Bipolar technology for VLSI circuits," *IEDM Dig.*, p. 63, 1983.

45. A. R. Alvarez, P. Meller, and B. Tien, "2 micron merged bipolar-CMOS technology," *IEDM Dig.*, p. 761, 1984.

46. T. Ikeda, T. Nagano, N. Momma, K. Miyata, H. Higuchi, M. Odaka, and K. Ogiue, "Advanced BiCMOS technology for high speed VLSI," *IEDM Dig.*, p. 408, 1986.

47. R. A. Kertis, D. D. Smith, and T. L. Bowman, "A 12 ns 256K BiCMOS SRAM,"

ISSCC Digest, p. 186, 1988; also in *IEEE J. Solid-State Circuits*, vol. 23, p. 1048, 1988.

48. H. V. Tran, D. B. Scott, P. K. Fung, R. H. Havemann, R. E. Eklund, T. E. Ham, R. A. Haken, and A. Shah, "An 8 ns Battery Back-up sibmicron BiCMOS 256K ECL SRAM," *ISSCC Digest*, p. 188, 1988.

49. N. Tamba, S. Miyaoka, M. Odaka, K. Ogiue, K. Yamada, T. Ikeda, M. Hirao, H. Higuchi, and H. Uchida, "An 8 ns 256K BiCMOS RAM," *ISSCC Digest*, p. 184, 1988.

50. N. Homma, K. Yamaguchi, H. Nanbu, K. Kanetani, Y. Nishioka, A. Uchida, and K. Ogiue, "A 3.5-ns, 2-W, 20-mm², 16-Kbit ECL bipolar RAM," *IEEE J. Solid State Circuits*, vol. SC-21, p. 675, 1988.

51. Y. Okajima, K. Toyoda, T. Awaya, K. Tanka, and Y. Nakamura, "64Kb ECL RAM with redundancy," *ISSCC Digest*, p. 48, 1985.

52. C-T. Chuang, D. D. Tang, G. P. Li, E. Hackbarth and R. R. Boedeker, "A 1.0-ns 5-kbit ECL RAM," *IEEE J. Solid State Circuits*, vol. SC-21, p. 670, 1986.

53. K. Sasaki, S. Hanamura, K. Ueda, T. Oono, O. Minato, K. Nishimura, Y. Sakai, S. Meguro, M. Tsunematsu, T. Masuhara, M. Kubotera, H. Toyoshima, "A 15 ns 1 Mb CMOS SRAM," *ISSCC Digest*, p. 174, 1988.

54. H. Shimada, Y. Tange, K. Tanimoto, M. Shiraishi, N. Suzuki, and T. Nomura, "An 18 ns 1 Mb CMOS SRAM," *ISSCC Digest*, p. 176, 1988.

55. F. List, S. Bell, S. Chu, J. Dikken, C. Hartgring, J. Raemaekers, B. Walsh, and R. Salters, "A 25 ns low-power full-CMOS 1 Mb SRAM," *ISSCC Digest*, p. 178, 1988.

56. H. Lee, B. El-Kareh, R. Glaker, G. Gravenities, R. Lipa, J. Maslack, J. Pessetto, W. Pokorny, M. Roberge, T. Williams, H. Zelelr, and K. Beilstein, "An experimental 1 Mb CMOS SRAM with configurable organization and operation," *ISSCC Digest*, p. 180, 1988.

57. H. De Los Santos and B. Hoefflinger, "On the analysis and design of CMOS-Bipolar SRAM's," *IEEE J. of Solid-State Circuits*, vol. SC-22, p. 616, 1987.

58. T. S. Yang, M. A. Horowitz, and B. A. Wooley, "A 4 nsec 4K × 1 bit two-port BiCMOS SRAM," *IEEE J. Solid State Circuits*, vol. SC-23, p. 1030, 1988.

59. A. R. Alvarez and D. W. Schucker, "BiCMOS technology for semi-custom integrated circuits," in the Proc. of Custom Integrated Circuits Conf., p. 22.1.1, 1988.

60. D. A. Hodges, P. R. Gray, and R. W. Brodersen, "Potential of MOS technologies for analog integrated circuits," *IEEE J. of Solid State Circuits*, vol. SC-13, p. 285, 1988.

61. A. Wieder, "Submicron bipolar technology—new chances for high speed applications," *IEDM Dig.*, p. 8, 1986.

EXERCISES

1. In a *p*-well CMOS, the *n*-type substrate has a doping concentration of 1×10^{15} cm^{-3} and the *p*-well concentration is 5×10^{16} cm^{-3}; calculate source/drain junction

capacitances and body effect coefficients for both *n*- and *p*-channel FETs assuming the gate oxide is 200 Å for both devices.

2. In a standard CMOS process, specify the necessary masking sequence in correct order and also state the number of masks needed for a double-level metal process.

3. To design a retrograde *p*-well with a well depth of 1 μm, how high an implant energy must be selected for the boron implant and what should be the minimum resist thickness if a resist mask is used during the high energy implant?

4. Repeat Exercise 3 and assume a retrograde *n*-well is needed.

5. Describe the major process steps and masking sequence for a CMOS/SOS and compare them with a bulk CMOS process.

6. In an *npn* bipolar transistor, the emitter sheet resistance is 50 $\Omega/\square$; the intrinsic base sheet resistance is 5000 $\Omega/\square$; the base width is 0.1 μm; the electron diffusion length in the base region is 10 μm: and, the electron mobility is 1000 cm^2/V-sec; calculate the common emitter current gain (β) and maximum frequency ($1/\tau_t$).

7. From the Gummel plot shown in Fig. 5.28, calculate the bipolar transconductance (g_m) defined as $\partial I_c/\partial V_{BE}$ and compare it with the transconductance ($g_m = \partial I_d/\partial V_g$) of the *n*-channel MOSFET shown in Fig. 2.13.

8. Describe the additional masking layers and process steps needed for adding *npn* bipolar transistors to an *n*-well CMOS process.

6

CMOS TRANSISTOR DESIGN

This chapter discusses transistor design for both n- and p-channel MOSFETs in a CMOS IC. Because this book is for VLSI technology, the design of transistors with micron and submicron channel lengths will be the aim of this chapter. Based on the device physics described in Chapter 2, an understanding of short-channel transistors will be gained by studying MOSFET scaling, then discussing non-scalable device parameters. This understanding will help to design short-channel FETs.

To design reliable devices, hot carrier effects must be taken into account. These effects are extremely important, especially for nMOS design because of the high impact ionization rate of electrons. Several sections will discuss this problem and its cure through device design.

For pMOS design, another problem exists because the pMOS transistors in a CMOS IC commonly pertain to buried-channel characteristics. This problem and its remedies will also be discussed. Finally, pMOS long-term reliability, which can also be a serious concern at submicron dimensions, will be studied.

TABLE 6.1 MOSFET Constant Electric Field
Scaling Laws

Parameter	Scaling Factor
Device dimensions t_{ox}, L, W, x_j	$1/S$
Doping concentration N_A	S
Voltage V	$1/S$
Current I	$1/S$
Capacitance $\varepsilon A/t_{ox}$	$1/S$
Delay time VC/I	$1/S$
Power dissipation VI	$1/S^2$
Power density VI/A	1
Power-delay product	$1/S^3$

6.1 MOSFET SCALING

For the past decade, MOSFET has been scaled down for higher density and higher speed. Based on MOSFET physics, the simple I-V relationships stated in Eqs. (2.24) and (2.29) basically remain when all device dimensions are scaled down by a common factor S where $S > 1$. Table 6.1 shows the changes in MOSFET dimensions and parameters as constant-field scaling[1] is applied. In this scaling, carrier behaviors such as velocity and ionization rate are maintained by keeping the electric field constant as device dimensions are reduced. Obviously, terminal voltages must also be reduced accordingly. To scale depletion width, doping concentration must be increased proportionally to keep a constant electric field. Other parameters are derived from Eqs. (2.24). In addition to the obvious area reduction, performance improvements are noticed in delay time and power dissipation. The major disadvantage of this scaling law is the reduction of supply voltage, which is not compatible with circuit requirements such as noise margin, current driving ability, and TTL compatibility.

Maintaining constant voltage while all device dimensions are scaled is another means of scaling MOSFET dimensions. Constant-voltage scaling solves the circuit incompatibility problem, however at the expense of a higher electrical field inside the device. The increasing E-field causes mobility degradation, hot carrier effects, oxide tunneling, and other reliability problems. A compromise between circuit performance and device reliability is proposed by the use of quasi constant-voltage scaling.[2] This approach scales voltage by $\sqrt{S}$ so that oxide field growth is no greater than $\sqrt{S}$.

6.2 NON-SCALABLE DEVICE PARAMETERS

In spite of the success of MOS scaling in the past, scaling limits will soon be approached as device dimensions decrease below one micron. The validity of extrapolating simple one-dimensional MOSFET equations to the submicron regime is questionable. Certain device parameters do not scale with the scaling of dimensions and doping concentration. Examples are threshold voltage, Fowler-Nordheim tunneling,[3] punchthrough voltage, and high-field mobility.

Threshold voltage.   According to Eq. (2.13), the threshold voltage of a MOSFET can be written in terms of device dimension such as oxide thickness (t_{ox}) and doping concentration (N_A):

$$V_t = \Phi_{ms} - Q_{fc}t_{ox}/\varepsilon_{ox} + 2(t_{ox}/\varepsilon_{ox})[kTN_A\varepsilon_s\ln(N_A/n_i)]^{1/2}$$
$$+ 2(kT/q)\ln(N_A/n_i) \qquad (6.1)$$

It is clear that V_t is a complex function of t_{ox} and N_A; thus, the simple scaling law does not apply here. Moreover, if the short- and narrow-channel effects described in Chapter 2 are considered, the threshold voltage is complicated even more through the dependence on channel length (see Eq. 2.45) and width.

Fowler-Nordheim tunneling.   Fowler-Nordheim tunneling through a thin gate oxide depends on electrical field across the oxide (E_{ox}) as follows[3]

$$I \propto WLE_{ox}^2\exp(-B/E_{ox}) \qquad (6.2)$$

E_{ox} is approximately equal to V_g/t_{ox} and B is proportional to $\Phi_b^{3/2}$, where Φ_b is the potential barrier height at silicon/oxide interface. For constant voltage scaling, the tunneling current increases rapidly with the electric field, thus placing a limit on minimum oxide thickness.

Subthreshold behavior.   In general, subthreshold swing (S_t) does not scale down with device dimension. At first-order approximation, the S_t value remains more or less constant according to Eq. 2.36. Thus, threshold reduction (necessary in constant-field scaling) shifts the $\ln I_d$ vs. V_g curve toward a lower V_g, resulting in higher off-state leakage for "scaled" transistors. The $I_d - V_g$ shift is even larger at higher drain bias due to drain-induced potential barrier lowering,[4] unless more sophisticated scaling guidelines are applied. An empirical relation has been found between various device dimensions for which long-channel subthreshold behavior has been maintained.[5] The proposed relation is

$$L_{min} = A[x_j t_{ox}(w_s + w_d)^2]^{1/3} \qquad (6.3)$$

where L_{min} is the minimum channel length for which long-channel subthreshold behavior is maintained. A is a proportionality constant, x_j is junction depth, t_{ox} is oxide thickness, and $w_s + w_d$ is the sum of source and drain depletion widths.

Punchthrough voltage.   Punchthrough current was described for a short-channel MOSFET operated under a subthreshold condition. It is often convenient to use punchthrough voltage (V_{pt}) as a device parameter in device scaling and device design. Punchthrough voltage is defined as the drain voltage at which a small but finite amount of drain current is observed at or near zero gate voltage. The amount of drain current used in punchthrough definition has not been universal. Depending on circuit requirement, this current level may vary from 1 pA for a DRAM circuit to 1 nA for a static circuit. As discussed in Chapter 3, Eq. 3.21 gives the following I-V relation when punchthrough occurs

$$I_d = -qD_n W(Z^*/L^*)(n_i^2/N_A)\exp[q(\Phi^* - V_s)/kT]$$

$$\{1 - \exp - [q(V_d - V_s)/kT]\} \qquad (6.4)$$

It is clear that punchthrough voltage is a complicated function of device dimensions and doping concentration, hence it does not follow the simple scaling laws mentioned above. In general, punchthrough voltage decreases during MOS scaling, again because of the drain induced barrier lowering effect.

Mobility degradation.   As discussed in Chapters 2 and 3, mobility decreases as an electrical field is increased. This mobility degradation expressed in Eq. (3.29) indicates that mobility decreases with device dimensions for constant-voltage or quasi constant-voltage scaling. As a result, transistor current is less than expected from simple scaling results. In fact, transistor current is even less because parasitic resistances do not scale with device dimensions.

Parasitic resistances.   As described in Chapter 2, Sec. 2.3.2, both transconductance (g_m) and channel conductance (g_d) suffer from the parasitic resistance associated with the source and drain regions. This parasitic resistance consists of the n^+ or p^+ S/D series resistance and the metal-to-Si contact resistance. It was pointed out[6] that both resistances increased as the junction depth was reduced during MOS scaling. In a conventional technology, the need for shallower junctions often leads to a lower n^+ or p^+ surface concentration, hence higher contact resistivity and higher sheet resistance. As a result, transconductance can actually decrease as gate length is reduced below one-half micron. However, this instance is not necessarily true when new technologies such as RTA (Rapid Thermal Annealing), salicidation (self-aligned S/D silicidation), and/or refractory metallization are used to produce shallow junctions with high surface concentration and improved metal-Si metallurgy. Supposing silicidation is in place, S/D sheet resistance as low as $2-3$ $\Omega/\square$ can be achieved, allowing MOSFET scaling down to a few tenths-micron, without suffering from associated parasitic resistance. A theoretical limit for contact resistance is around 10^{-8} Ω-cm^2, which is not yet a concern for MOSFET scaling.[7]

The maximum contact resistivity allowed in MOS scaling[8] has been analyzed. In the analysis, contact resistance increased by S^n for a given contact resistivity. The value n varied between 1 and 2 depending on the type of contact resistance.[9]

Assuming contact resistance scales at S^2 for the worst-case consideration, the corresponding result is slightly more stringent than what is required for actual devices. The intrinsic conductance of a FET (either channel conductance or transconductance) increases by S at most for constant voltage scaling, and stays unchanged for constant field scaling. For a fixed contact resistivity, the ratio of the contact resistance to the FET intrinsic resistance therefore increases to S^2 or S^3, depending on constant field or constant voltage scaling. To maintain negligible parasitic resistance, this ratio, which presents the relative contribution of the contact resistance to the device resistance, should be kept very small, say 1%. The only way to keep this ratio at a fixed low value during constant field or constant voltage scaling is to reduce contact resistivity to S^2 or S^3.

To illustrate this effect quantitatively, calculate the channel conductance of a 4-μm-long FET assuming 1000-Å gate oxide, 4-μm channel width (W/L = 1), 5-V power supply with 1-V threshold voltage (V_t), and 500-cm$^2/V \cdot$ sec electron mobility (μ_n). The calculated channel conductance (g_d), where $g_d = \mu_n C_{ox} W/L(V_g - V_t)$, for this device is 0.069 mS. The channel resistance R_d is therefore 14.47 kΩ. According to the criterion ($R_c/R_d < 1\%$), the contact resistance R_c must be <144.7 Ω, corresponding to a contact resistivity (ρ_c) <2.3 $\times$ 10^{-5} $\Omega \cdot$ cm^2 for a 4 μm-by-4 μm contact area. It has been commonly accepted that constant voltage (5-V supply) should be used for scaling MOSFETs from 4 to 2 μm. Scaling to 1 μm and below, a constant field scaling is assumed to avoid the high field problems such as mobility degradation, hot electrons, and electron tunneling. Using this combined scaling method as an example, the maximum contact resistivity for devices with 0.25 μm $\leq$ L $\leq$ 4 μm and W/L = 1 were calculated. For a device with W/L = 10, R_c was still no more than 10% of R_d assuming the contact size was maintained at L^2.

The results are shown in Table 6.2. If the contact resistivity is reduced below the required value shown in Table 6.2, the associated contact resistance does not result in a significant parasitic effect for the MOS device. Contact

TABLE 6.2 Maximum Contact Resistivity Allowed in MOSFETs (Ref. 8, © 1983 IEEE).

FET Channel Length (L),† μm	Contact Size, μm²	Contact Resistivity* (ρ_c), Ω-cm²	Power Supply (V_{DD}), V	Gate Oxide Thickness, Å	Threshold Voltage, V
4	4 × 4	2.3×10^{-5}	5	1000	1
2	2 × 2	2.9×10^{-6}	5	500	1
1	1 × 1	7.3×10^{-7}	2.5	250	0.5
0.5	0.5 × 0.5	1.8×10^{-7}	1.25	125	0.25
0.25	0.25 × 0.25	4.5×10^{-8}	0.63	63	0.13

*Maximum ρ_c values to keep $R_c/R_d \leq 1\%$
where R_c: contact resistance
       R_d: intrinsic device resistance
†$W = L$ for all the cases, W: channel width

resistivity in the order of 10^{-7} Ω-cm² is desirable for micron and submicron MOSFETs.

Similar analysis can be done to estimate the maximum S/D sheet resistance. Sheet resistance in the range of 40–100 Ω/□ is generally acceptable for a one-to-two micron CMOS. This range of resistance values can be achieved by using conventional n^+ and p^+ junctions. For submicron devices, silicided S/D junctions with the sheet resistance of only few Ω/□, are desired. Consequently, resistance at the contact hole depends on metal-to-silicide contact resistivity which is generally an order of magnitude lower than the resistivity of a metal-to-Si contact.

Parasitic capacitances.   As discussed in Chapter 3, the parasitic capacitances existing in a MOSFET are overlapping capacitance (C_{gdo} and C_{gso}) and junction capacitance (C_{jtot}). Both have been discussed. Even though the amount of S/D-to-gate overlap (l_{ov}) should scale with x_j and hence with channel length (L) in principle, it has not been the case in practice due to process constraints. Furthermore, for device reliability needs, a certain fixed amount of S/D-to-gate overlap is required. The details of this reliability issue will be discussed in subsequent sections. Also, the fringing field capacitance (C_{fr}) for S/D to gate does not scale. Consequently the ratio of overlapping capacitance to total gate capacitance usually increases during MOS scaling. Scaling of intrinsic capacitances (C_{gd} and C_{gs} for gate-to-drain and gate-to-source capacitance) is more complicated. In general, C_{gd} increases and C_{gs} decreases as MOSFET channel length is reduced.[10]

Interconnect scaling.   Despite the emphasis on transistor scaling in this section, interconnect scaling and its interaction with transistor scaling are important too.   Interconnect scaling for local interconnects and long-distance interconnects has been investigated.[11]   Local interconnects deal with interconnects between neighboring logic gates and scale with S.   Long-distance interconnects involve connection from one corner of a die to other corners. It is therefore appropriate to define a chip scaling factor of S_c.   Table 6.3 summarizes the results of interconnect scaling for an ideal case,[11] e.g., fringing field effects are not considered in capacitance scaling.   Notice that the delay time expressed by the RC constant remains unchanged for the local interconnect scaling or increases by $(S_c S)^2$ for the long-distance interconnect.   The

TABLE 6.3   (a) Local and (b) Long Distance Interconnect Scaling Rules (Ref. 11)

		"Ideal"
Length	L	$\rightarrow L/S$
Width	W	$\rightarrow W/S$
Thickness	H	$\rightarrow H/S$
Spacing	Z	$\rightarrow Z/S$
Insulator thickness	D	$\rightarrow D/S$
Resistance	R	$\rightarrow SR$
Capacitance (line-to-substrate)	C_{LS}	$\rightarrow C_{LS}/S$
Capacitance (line-to-line)	C_{LL}	$\rightarrow C_{LL}/S$
Delay time	RC	$\rightarrow RC$
Voltage drop	IR	$\rightarrow IR$
Current density	J	$\rightarrow SJ$
Contact resistance	R_C	$\rightarrow S^2 R_C$

(a)

		"Ideal"
Length	L	$\rightarrow LS_C$
Width	W	$\rightarrow W/S$
Thickness	H	$\rightarrow H/S$
Spacing	Z	$\rightarrow Z/S$
Insulator thickness	D	$\rightarrow D/S$
Resistance	R	$\rightarrow S^2 S_C R$
Capacitance (line-to-substrate)	C_{LS}	$\rightarrow S_C C_{LS}$
Capacitance (line-to-line)	C_{LL}	$\rightarrow S_C C_{LL}$
Delay time	RC	$\rightarrow S^2 S_C^2 RC$

(b)

INTERCONNECT LENGTH=0.5 cm

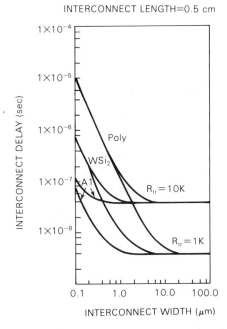

Figure 6.1 Long distance interconnect delay time versus interconnect width for three different interconnect materials and two transistor types (Ref. 11).

delay time of a long-distance interconnect driven by a CMOS inverter versus interconnect width has been computed for two driver transistor resistances (R_{tr}) and three interconnect materials. As shown in Fig. 6.1, the interconnect delay (t_{int}) increases when the interconnect width is reduced below a knee point. To stay in the flat region where t_{int} is minimal, the interconnect resistance (R_{int}) should be kept below 2.3 R_{tr}. For materials with higher sheet resistance such as polysilicon and silicide, the scaling of interconnect width is limited. Even for aluminum, scaling interconnect width down to submicron range can greatly increase signal propagation delay.

Electromigration, a current-induced mass transport, has produced a reliability limit in transistor and interconnect scaling. The median time to failure is

$$MTTF \propto wJ^{-2}\exp(E_a/kT) \tag{6.5}$$

where w is metal linewidth, J is current density and E_a is the activation energy. Because J increases by S or S^3 depending on constant-field or constant-voltage scaling, $MTTF$ degrades in the range of S^{-3} to S^{-7} according to Eq. 6.5. This strong dependence certainly places a scaling limitation on device reliability.[12] Other device reliability concerns including oxide wear-out and hot electron effects. These concerns will be discussed in subsequent sections.

6.3 TRANSISTOR DESIGN FOR SHORT-CHANNEL MOSFETS

As shown in Fig. 3.1, a device designer must have an understanding of the need for circuit performance and the limit of process capability. A general process architecture, e.g., the choice of well type in CMOS, should also be kept in mind. Device physics and device scaling form the necessary knowledge base. Various modelling tools are also required to deal with complex 2-D effects. Process simulation and device modelling are the core of device design. However, calculated results must be verified by experimental data. Several iterations are often needed to reach optimum conditions. Test structures are designed and fabricated to extract device parameters and establish electrical design rules. These outputs are used for circuit design and simulation.

For the past decade, 5 V supply has been the industry standard and will continue to be at least down to one-micron MOS technology. It is possible that a new standard supply such as 3.3 V or lower will be accepted for submicron MOS technology. Anyway, problems caused by the use of a fixed 5 volt supply on very small devices are discussed. The relevant device parameters are: threshold voltage, punchthrough voltage, transconductance, subthreshold characteristics, and parasitics. Junction capacitance and body effect are also important device considerations for circuit performance. This section will discuss device design through modelling, realization through process, and their impacts on circuit performance.

Based on simple scaling law, a 1-μm MOSFET should have approximately 200 Å of gate oxide, 0.2–0.25 μm junctions, and 2–3 $\times$ 10^{16} cm^{-3} substrate doping density. These device parameters also agree with the empirical relation expressed in Eq. 6.3. The 1-μm mentioned here represents the physical gate length on a finished wafer. Due to processing variation at lithography and etching, the gate length is commonly drawn slightly larger, say 1.2–1.3 μm on the mask to guarantee a physical gate length of 1 μm. The effective (or electric) channel length (L_{eff}), defined as the separation between source and drain junctions, is then approximately 0.7–0.8 μm due to S/D lateral diffusion. Reasonably good yield has been established for 200 Å oxide although some manufacturers use a slightly thicker (250 Å) oxide for higher yield. For n-channel MOSFETs, 0.25 μm junctions with sufficiently low sheet resistance, such as 50 $\Omega/\square$, are normally achieved by using a slow diffuser such as arsenic implanted at 80 KeV for about 5 $\times$ 10^{15} cm^{-2}.

For p-channel devices, deeper (0.35 μm)junctions with roughly 100 $\Omega/\square$ sheet resistance are commonly observed from a standard CMOS process. Using a 50 KeV $BF_2{}^+$ (equivalent to 9 KeV B$^+$) implant followed by rapid thermal annealing (RTA) can produce 0.25 μm p$^+$ junctions with 80 $\Omega/\square$ sheet resistance.

Scaling up the entire substrate doping density for short-channel devices is inadequate because the higher the substrate doping, the higher the junction capacitance and body effect. As a result, circuit performance, including both speed and noise margin, suffers. The influence of body effect on circuit operation is discussed in Chapter 4. It is therefore desirable to restrict the increase of substrate doping only near the Si surface so that short-channel effects can be avoided at minimal expense to junction capacitance and body effect. The use of a shallow channel implant in a lightly-doped ($1-2 \times 10^{15}$ cm^{-3}) substrate would provide this selective doping, hence offering adequate threshold and punchthrough voltages and meanwhile maintaining low junction capacitance and body effect.

A single 20 KeV B^+ implant at about 9×10^{11} cm^{-2} can produce appropriate threshold and punchthrough voltages for MOSFETs with drawn gate lengths of 1.3 μm nominally operated at a 2.5 V power supply.[13,14] A similar process has also been applied to 1.25-μm NMOS[15] and CMOS[16] devices using 5 V supply. Although some short-channel effects are observed, threshold and punchthrough voltages are reasonable. Meanwhile the corresponding junction capacitance is less than 10% of the gate capacitance and the body effect coefficient is only $0.1-0.2$ V$^{1/2}$. However, for devices with gate lengths drawn at ~1 μm ($L_{eff} \cong 0.55$ μm), one implant does not sufficiently satisfy all requirements, especially if a high resistivity (>10 Ω-cm) substrate is used. Another deeper (70 KeV) B^+ implant at 6×10^{11} cm^{-2} can be added to further suppress punchthrough current with minimal perturbation on threshold.[17]

As-implanted and final boron profiles simulated using SUPREM, a process simulator discussed in Chapter 3, are shown in Figure 6.2a. The corresponding subthreshold characteristics shown in Figure 6.2b indicate adequate threshold voltage and low punchthrough current at 2.5 V drain bias. Again, it is more difficult if a 5 V supply is used. Figure 6.3a shows that punchthrough current in an *n*-channel MOSFET made with the double boron-implant scheme increases rapidly as V_d approaches 5 V.[18] This rapid increase is due to the expansion of the depletion region as large values of V_d push the punchthrough path down to the lightly doped substrate, causing additional punchthrough current. Figure 6.3b shows this effect analyzed using the GEMINI program, a 2-D device model presented in Chapter 3. This subject will be discussed thoroughly when buried *p*-channel devices are described in Sec. 6.5. Another limitation which may stop constant-voltage scaling even sooner is high-field induced hot-carrier effects.

6.4 HOT CARRIER EFFECTS AND *N*MOS DESIGN

A high electrical field in a short-channel device at 5 V operation causes hot-carrier generation and impact ionization. This problem is more serious for

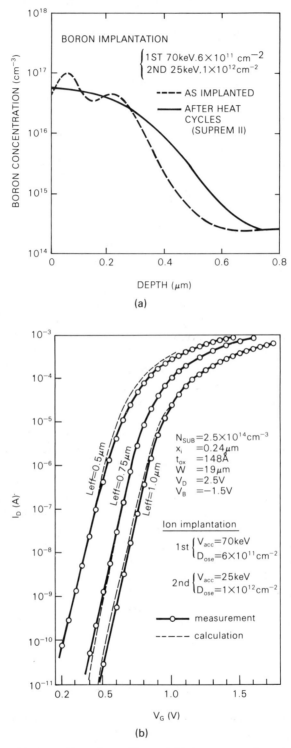

Figure 6.2 (a) Channel impurity profiles; and (b) the corresponding subthreshold characteristics for submicron MOSFETs (Ref. 17, © 1982 IEEE)

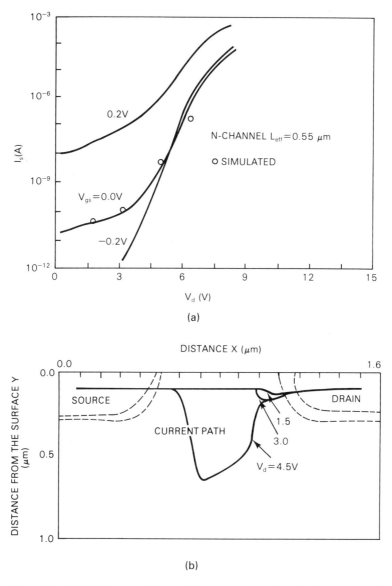

Figure 6.3   (a) I-V characteristics; and (b) current paths of a submicron *n*-channel MOSFET.

an *n*-channel MOSFET than for a *p*-channel device because, as described in Chapter 2, the impact ionization rate of electrons is one to two orders of magnitude higher than that of holes.   In fact, the hot carrier problem essentially does not exist in *p*-channel devices with $L_{\text{eff}} \cong 1$ μm.   Figure 6.4 shows that for a 1 μm MOSFET biased at $V_d = 5$ V, saturation current remains

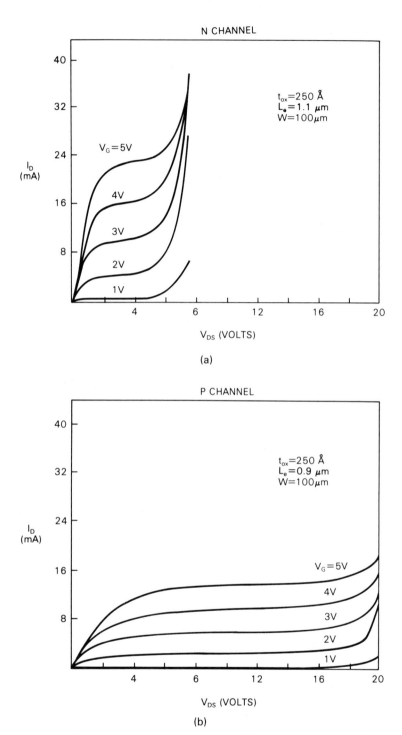

Figure 6.4 I-V characteristics of one-micron CMOS transistors: (a) *n*MOS and (b) *p*MOS.

constant for a *p*-channel device but increases substantially for an *n*-channel device due to impact ionization. Consequently, the device suffers from immediate damage and long-term deterioration. The immediate damage manifested in the form of substrate and gate currents is apparent. The long-term degradation arises from oxide instability caused by interface generation and/ or hot-carrier trapping.

6.4.1 Apparent Damage—Substrate and Gate Currents

A MOSFET acts like a resistor in linear operation. Voltage at the drain is low and uniformly spread across the channel. During saturated operation, the MOSFET channel is pinched-off and the drain voltage is high. Additional voltage dropped near the drain produces an exponential increase of channel potential as the drain is approached.[19] The channel potential near the drain can be approximated as

$$V(y) \cong V_{dsat} + V_o \exp(y/l) \tag{6.6}$$

where V_{dsat}, which was described in Chapter 2, represents the potential at the pinch-off or saturation point where y is set to zero (see the schematic in Fig. 6.5). V_o is a small constant and l is a characteristic length. The associated electric field is therefore

$$E(y) = |dV/dy| \cong [V(y) - V_{dsat}]/l \tag{6.7}$$

The maximum electric field is at the drain and can be written as

$$E_m = (V_d - V_{dsat})/l \tag{6.8}$$

Empirical relations between l and process parameters has indicated that

$$l \cong 0.22 \, t_{ox}^{1/3} x_j^{1/2} \tag{6.9}$$

where t_{ox} is the gate oxide thickness and x_j is the source/drain junction depth.[19,20] V_{dsat} depends on L_{eff} as follows[21]

$$V_{dsat} = \frac{(V_g - V_t)L_{eff}E_{sat}}{V_g - V_t + L_{eff}E_{sat}} \tag{6.10}$$

This expression reduces to Eq. (2.28) for large L_{eff} in a long-channel device. For very short-channel devices, $V_{dsat} \cong LE_{sat}$, implying that the *E*-field for the entire channel is at saturation. Between the extremes, for example $L_{eff} \cong 1 \, \mu$m, V_{dsat} is a function of L_{eff}, but is not directly proportional to L_{eff}. The dependence is weaker than a linear relationship especially at low V_g. During constant-voltage scaling, the increase of E_m comes directly from t_{ox} and x_j reduction, and indirectly from L_{eff} shortening.

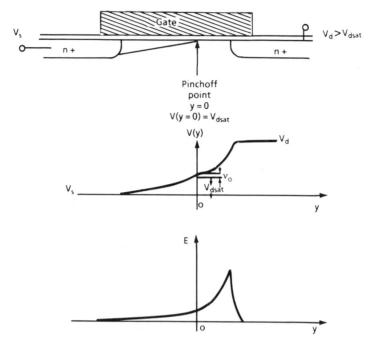

Figure 6.5 Schematics of channel potential and horizontal electric field of a MOSFET biased at saturation region.

Under the high field, electrons gain high enough energies and become hot. As shown in Fig. 6.6, these hot electrons can inject into the gate oxide, resulting in a gate current. They can also cause impact ionization near the drain and generate a hole current flowing into the substrate. As shown in Fig. 6.7a,[21] the measured substrate current, when plotted as a function of gate voltage, exhibits a bell-shape curve and increases with drain voltage. It peaks roughly at $V_g \cong V_t$ and decreases at lower V_g due to the lack of channel electrons for generating electron-hole pairs. It also decreases at higher V_g because the corresponding higher V_{dsat} lowers the E-field according to Eq. 6.8. If the same experimental data are plotted for the peak substrate current versus the inverse of $(V_d - V_{dsat})$, a straight line is obtained in a semi-logarithmic scale (Fig. 6.7b). Because $(V_d - V_{dsat})$ is directly proportional to electric field based on Eq. (6.8), the experimental observation in Fig. 6.7b fits well with the I_{sub} model commonly expressed as[19]

$$I_{sub} = C_1 I_d \exp - [\xi_i/(\lambda_e E_m)] \qquad (6.11)$$

where ξ_i is the threshold energy for impact ionization, λ_e is the hot-electron mean free path and C_1 is roughly a constant. So, $\lambda_e E_m$ is the amount of

Hot-electron effects in an n-channel MOSFET

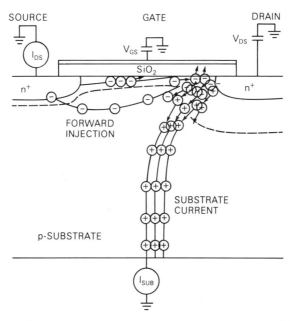

Figure 6.6   Schematic of hot-electron effects in an *n*-channel MOSFET.

energy gained by an electron before collision, and exp $- [\xi_i/(\lambda_e E_m)]$ is the probability that an electron will cause impact ionization.   The substrate current is undesirable because it causes substrate debiasing, snap-back breakdown, and latchup triggering.   It also induces photon emission, which generates minority carriers, which thereby degrade DRAM refresh time.[22]   In spite of these problems, substrate current is often measured for hot electron characterization because it is a good indicator of E_m, a very important parameter in hot electron effects.

The hot-electron induced gate current can also be modelled in a similar manner:

$$I_g = C_2(V_g, V_d)\, I_d \exp\, - [\xi_b/(\lambda_e E_m)]$$

or   (6.12)

$$= C_2(V_g, V_d)\, I_d \exp\, - [\xi_b/(kT_e)]$$

with the threshold energy ξ_b now corresponding to the Si/SiO$_2$ barrier energy, and T_e being the electron temperature.   Under the high field E_m, electrons with energy greater than ξ_b can inject into the gate oxide, thus constituting

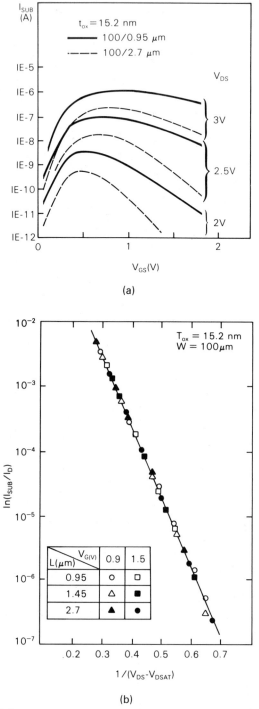

Figure 6.7 Substrate current as a function of: (a) gate voltage; and (b) drain voltage (Ref. 21; and © 1985 IEEE).

gate current. This simple equation is based on the lucky electron model[23,24] and $C_2 \approx 0.002$ for $V_g > V_d$. But, C_2 is actually a function of bias condition because it strongly depends on the electric field across the oxide (E_{ox}). For $V_g < V_d$, C_2 decreases rapidly because the corresponding E_{ox} acts as a retarding field to prevent injected electrons from reaching the gate. The E_{ox} tends to return electrons emitted into the oxide near the drain to the Si substrate.[25,26] To illustrate this phenomenon, the two-dimensional electric field distribution in the oxide has been calculated. Figure 6.8 shows the result obtained by using MINIMOS II.[26,27] Notice that the E_{ox} direction is reversed in Region 3. As a result, only part of the injected electrons contributes to actual gate current. Especially at low gate voltage, most injected electrons do not reach the gate. On the other hand, most ionization-induced hot holes, although at a much lower level, reach gate electrode when injecting into the oxide due to the aid of the E_{ox} field under the bias condition of $V_g < V_d$. The injection efficiency for holes is lower than that of electrons because the corresponding Si/SiO$_2$ barrier energy is higher, 4.8 eV instead of 3.1 eV for electrons.[27,28] However, the hot holes, when injected into the oxide, mostly contribute to hole gate current. Figure 6.9 shows the calculated gate currents and injected currents for both electrons and holes.[27] Notice the difference between the total injecting electron current and the electron gate current. Despite its use for EPROM programming, excessive gate current usually causes gate leakage and oxide breakdown. It can also induce long-term device degradation.

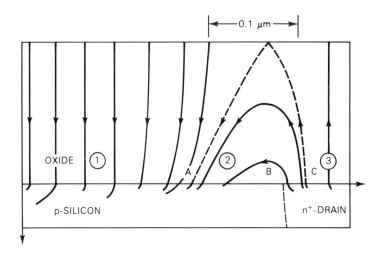

Figure 6.8 Two-dimensional electric field distribution near the drain of an *n*MOS. $V_d = 8$ V, $V_g = 7$ V, $V_B = -2.5$ V, $L_{eff} = 1.3$ μm and $t_{ox} = 400$ Å (Ref. 26, © 1984 IEEE).

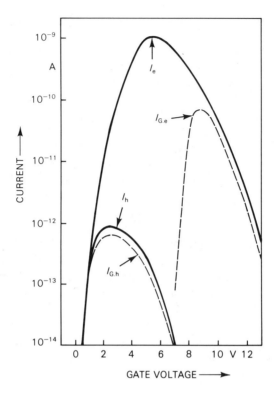

Figure 6.9 Calculated total emitted electron and hole current, I_e and I_h, as a function of gate voltage. $I_{G,e}$ is the electron gate current, and $I_{G,h}$ is the hole gate current. $V_d = 8$ V, $V_B = -2.5$ V, $W = 100$ µm, $L_{\text{eff}} = 1.8$ µm and $t_{ox} = 420$ Å (Ref. 26, © 1984 IEEE).

6.4.2 Long-Term Device Reliability

Hot carriers, when injected into oxide, can be trapped in the oxide or can generate interface states. The oxide traps, when occupied by one type of hot carrier, e.g., a hole, can subsequently convert themselves to electron traps. The hot carrier trapping and/or interface state generation changes threshold voltage, transconductance and the I-V relation in general. It is still not entirely clear whether electron trapping[28] or interface states generation[21] is primarily responsible for device degradation. But, usually a net negative charge density is observed after long-time stressing as evidenced by threshold increase. Hole trapping has been reported at low V_g,[26] but the effect is only secondary. Figure 6.10 shows typical $I_d - V_d$ characteristics before and after stressing.[29] Notice that drain current reduces after stressing and the reduction is more if the source and drain terminals are reversed after stressing. This reversal occurs because negative charges, when appearing near the source,

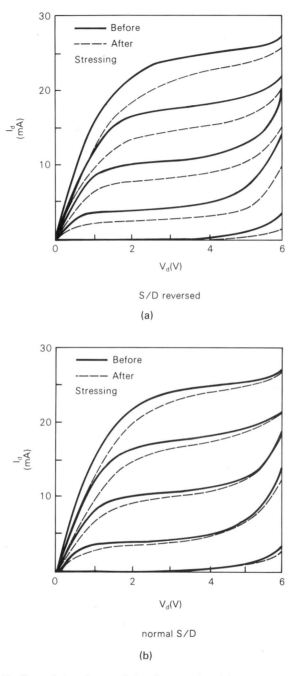

S/D reversed

(a)

normal S/D

(b)

Figure 6.10 Degradation characteristics after stressing: (a) source/drain reversed during post-stress measurement; (b) source/drain unchanged during post-stress measurement (Ref. 29, © 1984 IEEE).

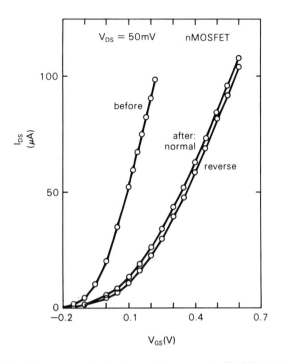

Figure 6.11   I-V characteristics before and after stressing (Ref. 21, © 1985 IEEE).

are more effective in reducing current.   This source/drain asymmetric effect is important for transistors used as transmission gates described in Chapter 4.   Figures 6.11 and 6.12 show typical $I_d - V_g$ characteristics before and after stressing.[21]   Threshold increase, transconductance decrease, and the increase of subthreshold swing constitute device degradation after stressing. Again the degradation is more severe when the source and drain are reversed after stressing.

Device lifetime (τ) can be defined by threshold voltage (V_t) change, transconductance (g_m) shift, or the variation of subthreshold swing (S_t).   In any case, lifetime follows I_{sub} in the following power-law relationship:

$$\tau = a(I_{sub})^{-b} \tag{6.13}$$

where a and b are constants.   This relationship, when plotted in a log–log scale, is shown in Figs. 6.13 and 6.14 for various devices biased at different drain voltages.[21]   It is not surprising to obtain the straight line relation because $\log(I_{sub})$ is a good indicator of peak electric field.   As the field increases, the lifetime should decrease.   Different symbols on the figure represent devices fabricated by using different technologies.   Notice that the slope of the log τ vs. log I_{sub} is roughly independent of technology; the actual lifetime value

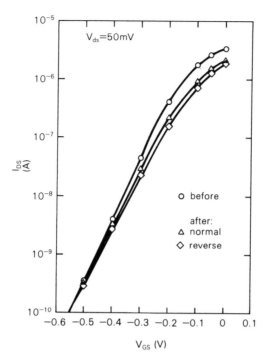

Figure 6.12   Subthreshold I-V before and after stressing (Ref. 21, © 1985 IEEE).

at a given I_{sub} may, however, vary by two orders of magnitude depending on process technologies.[21,30] In fact, the lifetime dependence on technology shown in Fig. 6.13 is stronger than its dependence on channel length variation (L).

Device lifetime in actual circuits.   Though most reliability work is done under static conditions, transistors in an actual circuit are stressed in a dynamic situation.   It is therefore important to understand device lifetime under dynamic stressing.   Modelling and characterization on simple circuits such as inverters, source followers, and pass gates has been done.[31]   It was found that switching speed (slew rate) is the most important parameter when generating substrate current.   Fast switching of an inverter's input voltage raises V_g above V_t when the output voltage is high.   This condition favors hot electron generation because the driver transistor is biased at $V_g > V_t$ and high V_d.   As shown in Fig. 6.15, a significant I_{sub} pulse appears in the input rising condition and its magnitude reduces as the rising is slowed.   Even at the highest slew rate, 100 ns in the experiment, substrate current peak is still lower than that observed in a corresponding static stress measurement.   Pulsed

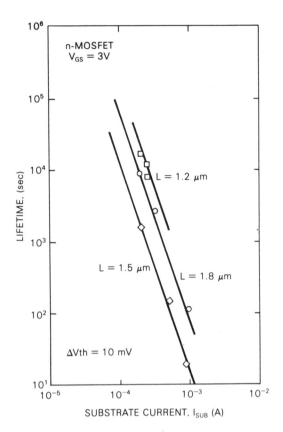

Figure 6.13 Lifetime as a function of substrate current, lifetime defined as 10 mV shift in threshold voltage (V_{th}) (Ref. 21, © 1985 IEEE).

measurements made on an inverter chain has provided the following simple correlation between static and dynamic lifetime (τ):

$$\tau_{\text{dynamic}} = \tau_{\text{static}}/(\text{duty cycle}) \qquad (6.14)$$

The experimental result is shown in Fig. 6.16.[32] This result allows extrapolation of static device lifetime to the actual device lifetime in a circuit if the duty cycle is estimated.

It was found that Eq. 6.14 needed to be modified when dynamic stress was done at high frequencies. Under a 25 MHz pulsed stress and an overlapping clocking for V_d and V_g, the transconductance degradation was increased over the corresponding static stress up to an order of magnitude as shown in Fig. 6.17.[33] This result was explained by a model in which charge trapping close to the interface was responsible for this enhanced dynamic degradation. Figure 6.18 shows the lifetime defined at 1% drain current reduction as a function of average number of $(I_{\text{sub}}^m/I_d^{m-1})^{-1}$, where $m = 2.9$.

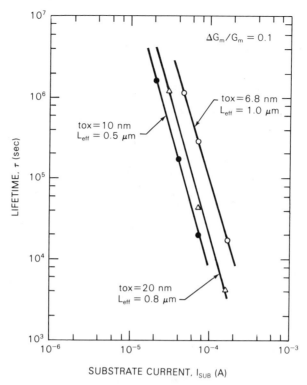

Figure 6.14 Lifetime as a function of substrate current, lifetime defined as 10% decrease in transconductance (G_m) (Ref. 21, © 1985 IEEE).

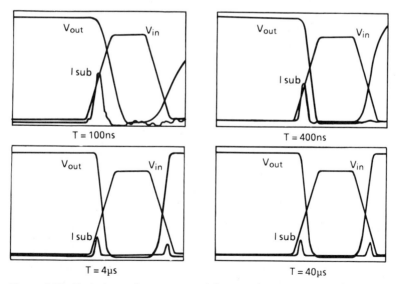

Figure 6.15 Typical waveforms measured from an inverter setup. The input transition times are 100 ns, 400 ns, 4 μs and 40 μs. The narrow pluses correspond to substrate currents (Ref. 31, © 1985 IEEE).

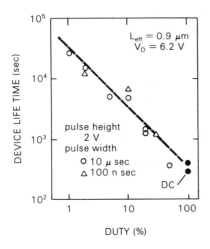

Figure 6.16 Lifetime versus duty cycle measured after dynamic stressing (Ref. 32, © 1986 IEEE).

The solid lines are calculated using the following generalized lifetime equation:

$$\tau_{\text{dynamic}} = C\langle I_{\text{sub}}^m / I_d^{m-1} \rangle^{-1} = C[1/T\int_o^T (I_{\text{sub}}^m / I_d^{m-1})dt]^{-1} \qquad (6.15)$$

A circuit reliability simulator, SCALE (Substrate Current And Lifetime Evaluator), was developed at the University of California–Berkeley to predict

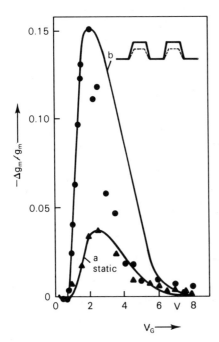

Figure 6.17 Experimental degradation results ($\Delta g_m/g_m$) versus gate voltage (V_G): dots and triangles—experimental data and solid curves—calculated results. Curve (a) represents the static results and curve (b) the dynamic case with the insert showing the V_D (full line) and V_G (dotted line) pulses with a 40 ns period (Ref. 33, © 1984 IEEE).

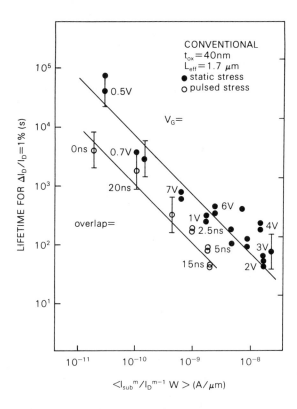

Figure 6.18 Static and dynamic degradation versus average substrate current with a constant or pulsed stress voltage of 8 V at drain. Different gate voltages are applied in the static case and the overlap of the gate and drain pulse was varied in the dynamic case (Ref. 33, © 1986 IEEE).

device lifetime in a circuit environment.[34]   This simulator was connected to SPICE in a pre- and post-processors configurations. The pre-processor takes input deck and BSIM parameters and requests a SPICE run to obtain the transient voltage waveforms at the drain, gate, source, and substrate of the user-selected device. The post-processor then calculates the transient substrate current based on the transient drain voltage and Eqs. 6.8 to 6.11. The drain current is obtained from the BSIM model and the lifetime is calculated by Eq. 6.15. Although the simulator is based on DC test data, its prediction on AC degradation provides a good estimation for most cases.

6.4.3 Drain Engineering for Reliability

Because high field near the drain is the main cause of hot carriers, the most effective way to mitigate the problem is to reduce the field by tailoring the doping profile near the drain. Lowering the doping concentration for the entire source/drain region would not be plausible because the associated series resistance would be too high, thus causing performance degradation as described in Sec. 6.2. Careful drain engineering must plan to reduce the doping level only in the periphery of the drain region. The implementation has been either to relax the one-sided abrupt junction profile or insert a couple of tenths-micron lightly doped region between the heavily doped drain and the channel.

Double-diffused drain (DDD). A double-diffused drain formed by two donor-type implants has been used to provide a graded rather than an abrupt junction. For n-channel MOSFETs, phosphorus (P) and arsenic (As) atoms are implanted into the source/drain region and subsequently diffused. Because phosphorus diffuses faster than arsenic, a graded junction is easily formed as shown in Fig. 6.19. Notice the lightly doped n^- rings around the n^+ regions. The transition from n^+ to n^- is actually not as abrupt as it appears in the figure. Figure 6.20 shows a typical impurity distribution along the channel direction of an nMOSFET.[35] The doping level drops by $2-3$ orders of magnitude from the n^+ to n^- drain regions. Doping profiles for conventional single-drain structures at two junction depths (x_j) are also plotted for comparison. The corresponding electric field distributions along the channel are shown in Fig. 6.21.[35] Notice the significant field reduction in the double-diffused case.

The DDD structure, although easily implemented, does result in a deeper junction, hence more short-channel effects and increased overlap capacitance. To reach an adequate trade-off between E-field reduction and acceptable short-channel effect, device/process optimization is necessary. Figure 6.22[35]

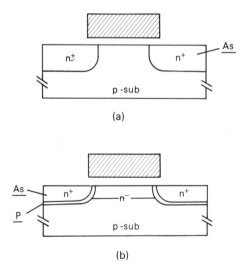

(a)

(b)

Figure 6.19 An nMOSFET with: (a) conventional single drain; and (b) double-diffused drain structure.

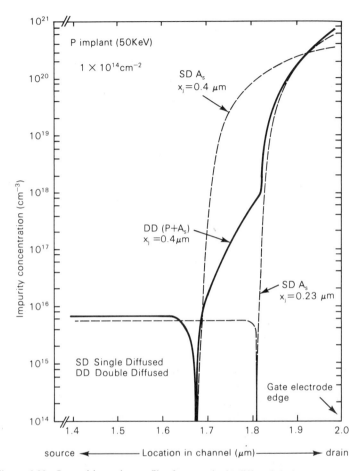

Figure 6.20   Lateral impurity profiles for two single-diffused drain structures and one double-diffused drain structure (Ref. 35, © 1985 IEEE).

shows the procedure used for designing an optimum *n*-channel DDD transistor at $L_g = 2$ μm. The requirement for minimal short-channel effects is set at <0.3 V threshold shift as L_g is reduced from 4 to 2 μm. Appropriate parameters such as oxide thickness, junction depth, and channel doses are then determined.   If the *S/D* sheet resistance is limited at <40 $\Omega/\square$ for maximum acceptable series resistance, junction depth set by the As implant is then fixed and so is the n^- diffusion distance, Δx_j. Process conditions such as As annealing and P implant can then be decided. Finally, optimal P dose and subsequent annealing conditions are selected to minimize electric field at the stress voltages that correspond to the worst case. The optimized process/ device parameters for a 2 μm DDD structure are presented in Table 6.4.[35]

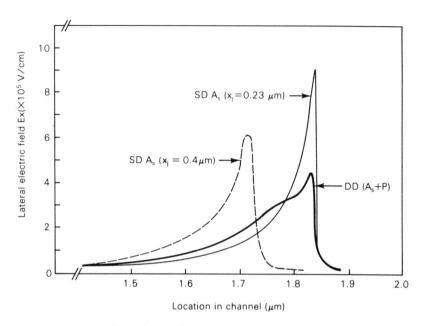

Figure 6.21 Lateral electric field distributions associated with the three structures shown in Fig. 6.20 (Ref. 35, © 1985 IEEE).

Lightly-doped drain (LDD).   The DDD structure can easily be implemented and is effective in reducing hot-carrier effects for 1.5–2.0 μm n-channel MOSFETs.   The extension of this structure to $\leqslant$1 μm transistors however is not ideal due to the inherently deeper S/D junctions, which cause increased short-channel effects and more gate-to-source/drain overlap capacitance.   Alternatively, another type of drain engineering has been developed to form an LDD (Lightly-Doped Drain) structure that uses a lightly-doped

TABLE 6.4 Process and Device Parameters for the Optimized Double-Diffused Structure (Ref. 35, © 1985 IEEE).

	Parameters	Conditions
1	Gate length	2 μm
2	Gate oxide thickness	350 Å
3	Channel implant dose	5×10^{11} cm^{-2}, 70 keV
4	Drain junction depth	0.4 μm
5	n^- diffusion region length	0.14 μm
6	P implant dose	1×10^{14} cm^{-2}, 100 keV
7	Annealing after P ion implantation	22 min at 1000°C
8	Annealing after As ion implantation	30 min at 950°C

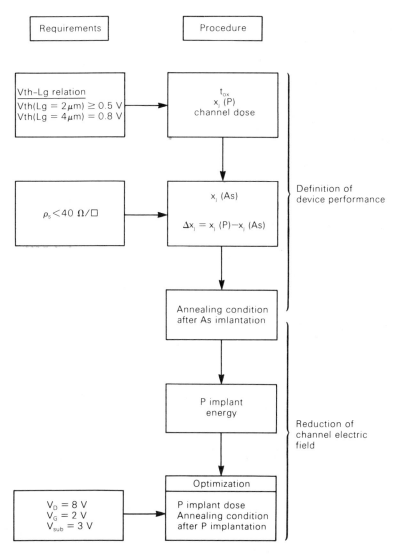

Figure 6.22   Optimization procedure for the double-diffused structure (Ref. 35, © 1985 IEEE).

buffer zone between the n^+ drain and the gate.   The associated process is more complicated than the DDD structure.   As shown in Fig. 6.23, a low-energy (40 KeV) phosphorus at the dose of $5-30 \times 10^{12}$ cm^{-2} is normally implanted to form the n^- LDD region after polysilicon gate delineation. Then, a CVD oxide is deposited and subsequently RIE etched to form a spacer at the sidewall of the polysilicon gate.   The oxide spacer then serves as a mask for the standard n^+ arsenic implant.

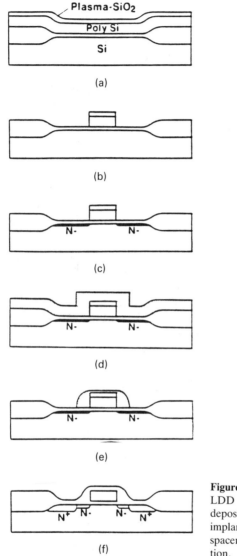

Figure 6.23 Process sequence for an LDD nMOS transistor. (a) after Poly deposition; (b) after Poly etch; (c) LDD implant; (d) spacer deposition; (e) spacer etch; (f) N$^+$ implant & passivation.

The LDD structure and associated E_m are shown in Fig. 6.24 and compared with a conventional FET.[36] The E_m in Eq. (6.8) is now lowered through the increase of the characteristic length l' as

$$E_m = (V_d - V_{dsat})/l' \qquad (6.16)$$

where $l' = l + L_{n-}$, and L_{n-} is the lightly doped n^- length. Consequently, I_g and I_{sub} are also reduced. Fig. 6.25 shows typical I_{sub} values for conven-

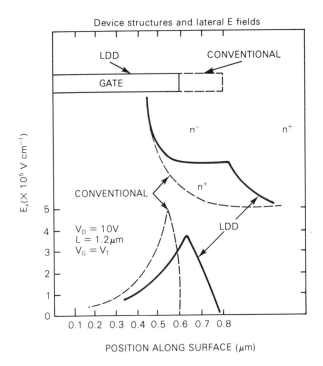

Figure 6.24 Device structures and lateral electric fields for a conventional and an LDD MOSFET (Ref. 36, © 1982 IEEE).

tional and LDD devices[37]; the lower the n^- dose, the more the I_{sub} reduction. However, the n^- region results in an extra resistance, giving lower g_m and I_d. As shown in Fig. 6.26,[38] a lower n^- dose causes more I_d reduction. Tradeoff between low substrate and high drain currents suggests that the I_{sub}/I_d ratio should be as low as possible. According to Eq. 6.11, this ratio is an indicator of the E_m value.

An LDD structure may shift the E_m location from under the gate into the n^- region, although this case is not necessarily true. When this shift occurs, negative charges are generated in the oxide above the n^- region during stress (Fig. 6.27). These charges tend to deplete the n^- region, especially if it is formed with low n^- dose. Hence, series resistance in the n^- region is further increased, thereby causing additional g_m degradation. A small but finite drain-to-gate overlap (≈ 0.1 μm) should be provided to ensure that the E_m location stays under the gate electrode.[39] E_m location with respect to gate edge is therefore very important in LDD device reliability. A quasi-2-D analytic approximation has been derived as[40]

$$\frac{dE_y}{dy} = \frac{v_g - v_{dsat}}{l^2} - \frac{qN_d(y)}{\varepsilon_{si}} \tag{6.17}$$

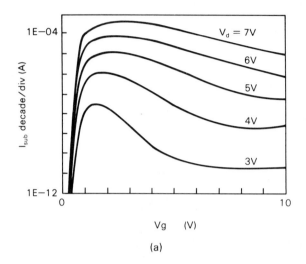

(a)

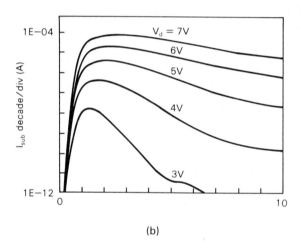

(b)

Figure 6.25 Substrate currents for: (a) a conventional; and (b) an LDD transistor (Ref. 37).

for the lateral field in the n^- region outside the gate. For a high n^- dose, the corresponding LDD concentration (N_d) is large. Therefore, dE_y/dy (the slope of the field) is negative outside the gate, implying the peak field is under the gate. This approximation compares well with 2-D numerical simulation performed using PISCES. Experimental results for LDD devices have also confirmed that an n^- dose at $1 - 2 \times 10^{13}$ cm^{-2} provides a long lifetime, even though the corresponding I_{sub} is not minimal at this dose. This

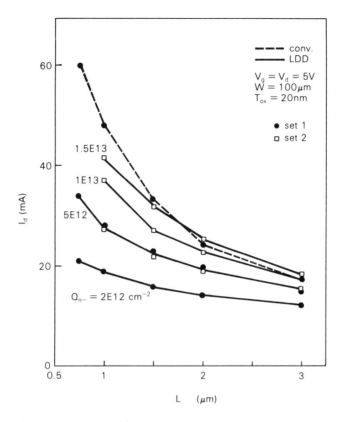

Figure 6.26   Drain current in LDD MOSFETs with various gate lengths (L) (Ref. 38, © 1984 IEEE).

relatively high dose also minimizes g_m degradation caused by additional series resistance.

Although n^- length is an important parameter, it is limited by process variations in polysilicon re-oxidation, oxide spacer formation, and subsequent n^- anneal. In general, a ~ 0.2 μm wide spacer is used in conjunction with a 0.3 μm thick polysilicon gate for process controllability. In a typical MOS process, subsequent to poly etching and n^+ implant, the gate oxide remaining above the S/D region is etched away because it is likely contaminated during RIE etching and damaged during high-dose implant. A fresh oxide is regrown on Si substrate as well as on and around the poly gate. This re-oxidation procedure as extended to an LDD process is commonly performed immediately after the n^- implant. The procedure has significant implication on n^- to poly-gate overlap. The schematics in Fig. 6.28 show this problem. Due to the increased oxidation rate on heavily-doped n^+ polysilicon, poly

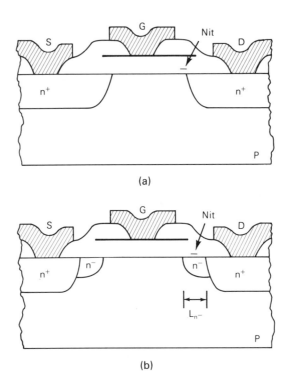

(a)

(b)

Figure 6.27 Degradation in: (a) a conventional MOSFET; and (b) an LDD MOS-FET.

retreat at the poly sidewall can result in a gap between the poly edge and the n^- region. Oxidation-induced encroachment at the poly edges also creates a thicker oxide there, producing a so-called Graded Oxide Gate (GGO) structure.[39] To ensure an adequate overlap, the subsequent n^- anneal must be sufficient to drive the n^- under the gate, more favorably beyond the GGO region. Because a shallow n^- junction is desirable for short-channel MOS-FETs, poly re-oxidation should be minimized to avoid excessive n^- anneal which would otherwise be needed. A short, dry oxidation is preferable to reduce the accelerated oxidation rate on n^+ polysilicon.

At present, 1 μm nMOS transistors can be made reliably with an adequate LDD structure and process. Figure 6.29 shows the extrapolated lifetime[37] defined as 10% reduction in linear transconductance. It represents the worst case for LDD MOSFETs. The best LDD transistors have an extrapolated lifetime of more than 10 years at a 5 V supply. They are fabricated with a 40 KeV phosphorus implant at 2×10^{13} cm^{-2} dose, a dry re-oxidation to form thin poly oxide and a sufficient n^- anneal to ensure proper n^- to poly overlap. The spacer width is 0.2 ± 0.05 μm and the n^+ is formed with standard arsenic S/D implant.

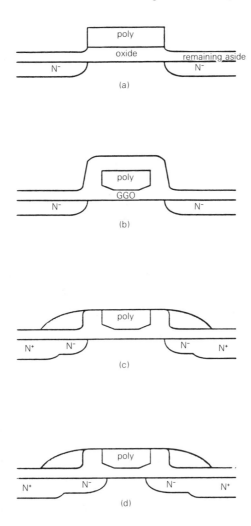

Figure 6.28   Schematics showing poly-gate retreat and graded-gate oxide (GGO). (a) after poly etch and N^- implant; (b) after reox; (c) after space formation and N^+ implant; and (d) after activation anneal.

The reliability problem can also be relaxed by circuit design. Lower duty cycle, limited voltage swing, clocked logic and/or NAND gate building blocks will ease the hot-electron problem through reducing either the drain voltage or the time that the device is under voltage stress.

The DDD scheme has a simpler process and suffers less transconductance degradation, but the LDD structure has fewer short-channel effects due to shallower S/D junctions.[41]   The lightly doped *n*-region in an LDD device is defined by the spacer width rather than by impurity diffusion in the DDD

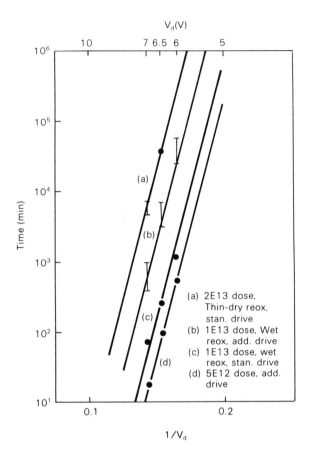

Figure 6.29 Lifetime for 1 μm LDD transistors fabricated with different processes (Ref. 37, © 1987 IEEE).

case, hence the n^- dimension in the LDD can be more tightly controlled. These reasons make LDD more extendable to MOSFETs with $L \cong 1$ μm. It can also be applied to submicron p-channel MOSFETs in which an LDD-like structure may become necessary if 5 V supply is required. Additional drain engineering is needed to make reliable submicron MOSFETs operate at 5 V. Optimization in device structure such as a combination of LDD and DDD in the form of a deeper n^- and a shallower n^+ with spacer has shown good, reliable results.[42] The formation of a p-pocket around the S/D region, called the Double-Implant (DI) LDD, is also effective in maximizing allowable drain voltage.[43] Buried-channel structure has been considered to reduce hot-electron effects because of its lower electric field and deeper current path.[44] Key processes such as gate oxide, spacer material and annealing condition[45] should also be refined to reduce hot carrier trapping and/or interface state generation.

6.5 BURIED-CHANNEL EFFECTS AND *p*MOS DESIGN

Chapter 2 introduced buried-channel MOSFETs. Because the *p*-channel transistors in a standard CMOS IC are buried-channel devices, it is important to understand buried-channel behavior prior to the discussion of *p*MOS transistor design.

6.5.1 Buried Channel Behavior

Threshold scaling problem.   A standard MOS process uses heavily doped *n*-type (n^+) polysilicon for the gate material. Although p^+ poly gates have been attempted, threshold instability caused by boron penetration has always been a problem. The n^+-poly is ideal for short *n*MOSFETs because the corresponding work function difference ($\mathbf{\Phi}_{ms}$ value) is desirable for scaling threshold voltage. As expressed in Eq. 2.13, threshold voltage consists of fixed oxide charge density (Q_{fc}), bulk-charge term (Q_B), and band-bending in Si ($2\,\mathbf{\Phi}_B$), in addition to the $\mathbf{\Phi}_{ms}$ term:

$$V_t = \mathbf{\Phi}_{ms} - Q_{fc}/C_{ox} + Q_B/C_{ox} + 2\,\mathbf{\Phi}_B \qquad (6.18)$$

where $Q_B = (2\varepsilon_s q N_A (2\mathbf{\Phi}_B))^{1/2}$, $\mathbf{\Phi}_B = (kT/q)\ln(N_A/n_i)$ and both increase as the channel doping level is raised for scaling down FET dimensions. The second term (Q_{fc}/C_{ox}) is negligible for state-of-the-art MOS process because of low oxide charge density and thin gate oxide. As shown in Fig. 3.30(a) for an *n*-channel device, the $\mathbf{\Phi}_{ms}$ for an n^+ poly-gate on a *p*-type substrate is

$$\mathbf{\Phi}_{ms}\,(n\text{-}ch) = \mathbf{\Phi}_m(n^+\text{-poly}) - \mathbf{\Phi}_{se}(p\text{-sub}) = -(Eg/2 + \mathbf{\Phi}_B) \qquad (6.19a)$$

with its value between -0.9 to -0.95 V for 10^{16} cm$^{-3} < N_A < 10^{17}$ cm^{-3}, the appropriate doping range for micron and submicron technologies. This negative value compensates the Q_B and the $2\,\mathbf{\Phi}_B$ terms and lowers the threshold to approximately 0.6 V, which is appropriate for VLSI. From Fig. 6.30(b), the *p*-channel $\mathbf{\Phi}_{ms}$ for an n^+ poly-gate on an *n*-type substrate is

$$\mathbf{\Phi}_{ms}\,(p\text{-}ch) = \mathbf{\Phi}_m(n^+\text{-poly}) - \mathbf{\Phi}_{se}(n\text{-sub}) = -(Eg/2 - \mathbf{\Phi}_B) \qquad (6.19b)$$

with its value between -0.2 to -0.15 V for the same doping range. Because all other terms in the V_t expression are negative for *p*MOS and again, the magnitudes of Q_B and $2\,\mathbf{\Phi}_B$ terms become larger for short-channel FETs, the resulting threshold voltage has a large negative value.[46,47] Remember, threshold voltage with large magnitude is undesirable because it degrades current driving ability. Figure 6.31 shows threshold scaling with respect to substrate doping and oxide thickness for an n^+-poly gate.[48]

Notice $V_t(n\text{-}ch) = 0.6$ V, but $V_t(p\text{-}ch) = -1.7$ V for $t_{ox} = 250$ Å and $N_A = 5 \times 10^{16}$ cm^{-3}. Physically, the work function difference associated with an n^+ poly-gate has different effects on *n*- and *p*-channel FETs. Figure

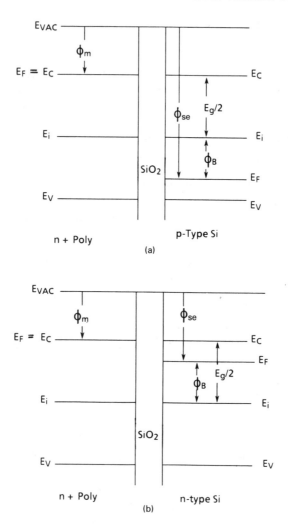

Figure 6.30   Band diagrams and work function differences for n^+ poly gate MOS structures: (a) p-type Si and (b) n-type Si substrate.

6.32 shows the band diagrams for p- and n-type substrates at equilibrium and inversion conditions.   Because the Fermi levels in the Si substrate and the n^+-poly gate must be aligned at thermal equilibrium, resultant band bending tends to promote inversion for the n-channel MOSFET but opposes inversion for the p-channel device.   Consequently, the gate voltage needed to invert the Si surface is small for the n-channel FET, but too large for the p-channel case.

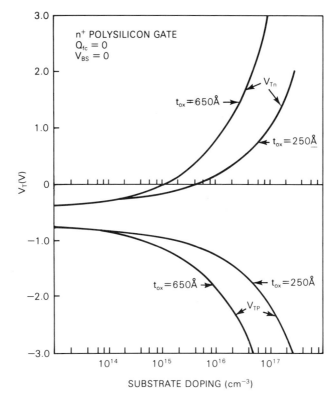

Figure 6.31   Threshold voltages versus substrate doping for *n*- and *p*-channel MOS-FETs using n^+ polysilicon gate.

Threshold adjust and buried-channel behavior.   Despite the threshold instability associated with a p^+ poly gate, if one is used in a CMOS process, the threshold scaling problem then appears in *n*-channel devices.   If differently doped poly gates are used for different device types, the associated process becomes much more complicated.   Refractory metals or metal silicides such as molybdenum have symmetrical Φ_{ms} for *n*MOS and *p*MOS, hence they can be used to compromise the performance of the two transistor types[49–51]. However, their compatibility with gate oxide and other MOS processes have not yet been fully demonstrated.   The present industry practice is to adjust the *p*MOS threshold by introducing a thin sheet of negative charges, such as ionized boron impurities, at the Si/SiO$_2$ interface.   The intent is to introduce a negative Q_{fc} term in Eq. 6.18 so that V_t can be less negative.   This practice is commonly performed with a boron implant which however, always results in a *p*-layer with finite thickness, as shown in Fig. 6.33 for two device configurations.   As a consequence, the channel potential is modulated; more

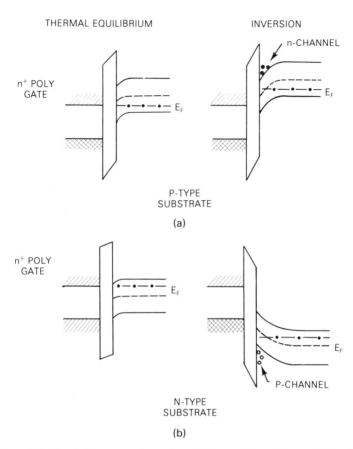

Figure 6.32 Band diagrams at thermal equilibrium and inversion conditions for both: (a) *p*- and (b) *n*-type substrates.

specifically, the channel potential minimum is moved away from the surface, causing buried-channel operation as described in Chapter 2.

Figure 6.34(a) shows the potential under the gate at the distance from the source where the longitudinal electric field vanishes.[52] The solid and dashed curves correspond to transistors made in the *n*-well and *n*-substrate, respectively. The potential profile of a surface-channel *n*MOS is also schematically plotted for illustration. Notice that the potential minimum is slightly away from the Si surface for both buried *p*-channel devices. Comparing the two *p*MOS devices, the potential valley is closer to the surface and the spread is less for the device made in the *n*-well because of the thinner *p*-layer and the higher background concentration (Fig. 6.34a).

When compared with a surface-channel device, buried-channel devices have poorer subthreshold characteristics due to spread of carriers and deeper carrier centroid.[52,53] Carriers, which are located farther away from the gate, are more difficult to control by gate voltage. The thinner the *p*-layer, the

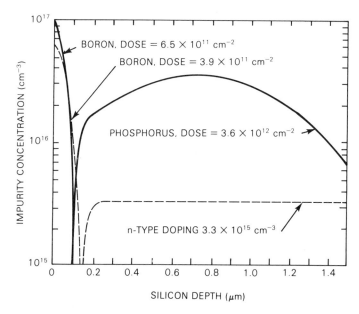

Figure 6.33 Simulated doping profiles in *p*MOS channel regions: dashed lines, lightly doped substrate; solid lines, retrograde *n*-well (Ref. 52, © 1982 IEEE).

better the subthreshold swing (smaller S_t value).   As shown in Fig. 6.34(b), finite leakage at zero gate bias is commonly observed for short *p*-channel devices, especially at a high drain bias.   Higher background concentration can improve this leakage because of the reduction of drain-induced barrier lowering (DIBL).[5]   The leakage current observed at high V_{ds} and $V_g < V_t$ is caused by punchthrough phenomenon due to DIBL.   Again, the subthreshold current of a surface-channel *n*MOS is plotted for comparison.

In a CMOS IC, punchthrough is a critical concern for *p*MOS FETs, even when they are made in *n*-wells, because of their associated buried-channel behavior and deeper p^+ S/D junctions.   The punchthrough current modelled by GEMINI is governed by Eq. (6.4) and the equation can be simplified for $V_{ds} \gg kT/q$, the only condition with practical interest.   The logarithmic form of the simplified equation is

$$\log(I_s) = A - q\Phi_b/kT \tag{6.20a}$$

where

$$A = \log[qDpW(Z^*/L^*)(n_i^2/N_A)] \tag{6.20b}$$

and

$$\Phi_b = \Phi^* - V_s \tag{6.20c}$$

Φ_b is the potential barrier height, Φ^* is the potential peak, and V_s is the source potential.   As shown in Fig. 6.35, Φ^* and hence Φ_b decrease as V_{ds} increases because of drain-induced barrier lowering.   Simulated results for

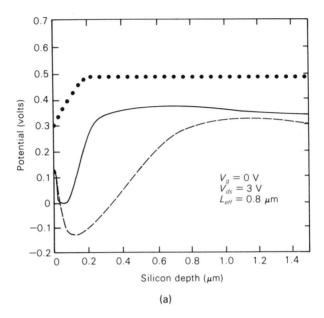

(a)

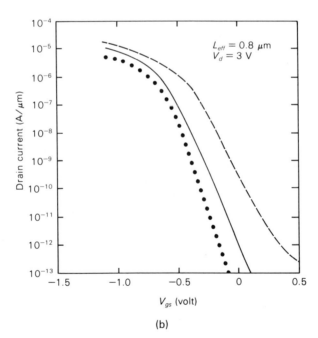

(b)

Figure 6.34 (a) Channel potential profiles; and (b) associated subthreshold I-V curves. Dashed lines: buried-channel pMOS in a lightly-doped substrate, solid lines: buried-channel pMOS in a retrograde n-well, dotted lines: surface-channel nMOS schematically drawn for comparison (Ref. 52, © 1982 IEEE).

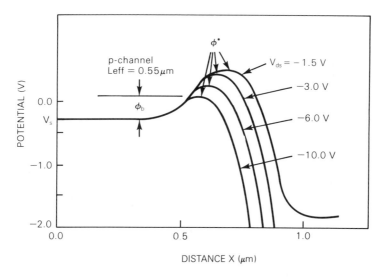

Figure 6.35 Potential distributions along the buried-channel of a *p*MOSFET for various drain voltages.

*p*MOSFETs in an *n*-well are shown in Fig. 6.36.   It is obvious that the barrier lowering is linearly proportional to the increase of drain voltage.   Eq. (6.20a) can therefore be written as

$$\log(I_s) = A - q(\Phi_{b0} - b|V_{ds}|)/kT \qquad (6.21a)$$

$$= A' - V_{ds}/\eta \qquad (6.21b)$$

where $A' = A - q\Phi_{b0}/kT$ and $\eta = kT/bq$. Φ_{b0} is the barrier height at $V_{ds} = 0$ and b represents the DIBL rate—both can be determined from Fig. 6.36.   A linear relationship is expected between I_s in logarithm and V_{ds}. Figure 6.37 shows the experimental data for a 0.8 μm *p*MOS transistor made in an *n*-well.[18]   The data agrees well with the simulated results.   The parameter, η, ($\eta = \partial V_{ds}/\partial \log(I_s)$) is the incremental voltage that the drain can sustain before I_S increases by another order of magnitude.   The larger the η value, the lower the DIBL rate, hence the better the punchthrough resistance in the MOSFET.   From Fig. 6.37, η is about -4.4 V/decade for $V_g = 0$ V.   This value means that if punchthrough current is 10 pA at $V_{ds} = -1$ V, the current will increase to 1 nA at $V_{ds} \cong -10$ V.   For VLSI devices, even a small amount of leakage may not be tolerated due to the large number of devices on a chip.   Considering a million-device chip, 0.1 μA leakage per device would generate a total of a few-tenths watt power dissipation.   For a DRAM with 50 fF charge storage, off-state leakage should be less than 1 pA, to ensure that the voltage at the storage node shifts less than 20 mV per msec.

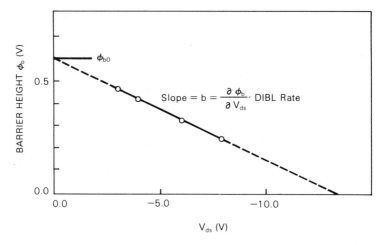

Figure 6.36 Potential barrier height as a function of drain voltage.

An obvious solution to the punchthrough problem is to make the p-type buried layer as thin as possible. The thinner the p-layer, the more the device resembles a surface-channel FET. Figure 6.38 shows the reduction of sub-threshold swing with thinner p-layer thickness (Y_j) for two different oxide thicknesses.[54] This reduction can be achieved by using a shallow BF_2^+ implant, higher background doping concentration, and/or a reduced thermal

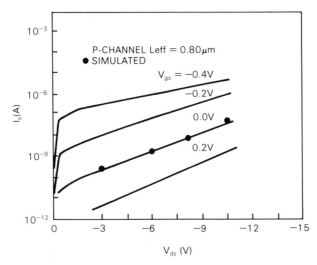

Figure 6.37 Subthreshold I-V characteristics versus drain voltage for a submicron pMOSFET.

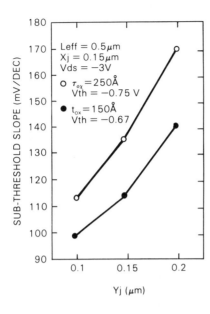

Figure 6.38 Simulated subthreshold swing versus buried-layer thickness (Y_j) for two oxide thicknesses (Ref. 54, © 1984 IEEE).

processing cycle to restrict boron diffusion. In an *n*-well or twin-tub architecture, a retrograde well profile and even a second *n*-type implant in the well[54-57] could be used to squeeze the *p*-layer closer to the Si surface. All of these process methods however, should not significantly upset other device parameters such as junction capacitance and body effect.

Another solution is to use a shallow *S/D* junction for reducing punchthrough and other short-channel effects. As shown in Fig. 6.39, both subthreshold swing and off-state leakage are significantly reduced when shallower p^+ junctions are used for *p*-channel source/drains. This method can be achieved again with BF_2^+ implant and/or rapid thermal annealing (RTA). It is difficult however to form a very shallow but still heavily doped p^+ junction even with the RTA technique. The LDD structure, which is commonly used to mitigate the hot-electron effect in *n*MOS, has been applied to half-micron *p*MOSFETs primarily to decrease punchthrough current at minimum expense to *S/D* resistance. Figure 6.40 shows that punchthrough voltage (V_{PT}), defined as V_{ds} at $V_g = 0$ V and $I_d = 1$ μA, is increased by 2–3 V when LDD is used.[55] In other words, for a specified punchthrough voltage, incorporating an LDD structure allows a shorter poly gate to be used. The LDD structure may result in a decrease in drain current due to the additional series resistance. Figure 6.41(a) and (b) shows punchthrough voltage and drain current as a function of poly gate length (L_g).[58] Notice that the V_{PT} curve shifts to the left indicating that for a given punchthrough voltage the poly gate can be shortened by approximately 0.2 μm. Meanwhile, the drain current reduces, but only marginally at larger L_gs. New device structures other than LDD have also been proposed to minimize punchthrough current. For example,

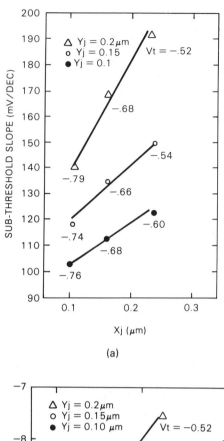

(a)

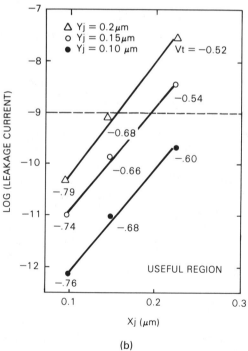

(b)

Figure 6.39 Simulated subthreshold characteristics: (a) subthreshold swing; and (b) off-state leakage current versus junction depth (x_j) for various buried-layer thicknesses. $L_{eff} = 0.5$ μm, $t_{ox} = 250$ Å and $V_{ds} = -3$ V (Ref. 54, © 1984 IEEE).

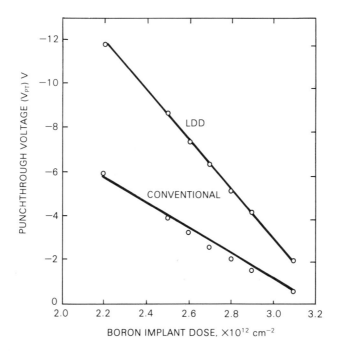

Figure 6.40   Punchthrough voltage as a function of boron implant dose for a 0.4 μm *p*MOSFET with and without LDD structure (Ref. 55, © 1986 IEEE).

a small segment of an *n*-type region can be inserted between the *p*-type *S/D* and the gate as a punchthrough stopper. This structure is often referred to as a halo structure and has been used in suppressing punchthrough current for half-micron FETs.[60] Furthermore, the LDD structure is effective in increasing submicron *p*MOS reliability.

6.5.2 *p*MOS Reliability

Previous discussion has indicated that the hot-carrier problem has not been serious in *p*MOSFETs because of the corresponding lower impact ionization rate. A secondary reason is that the buried-channel behavior associated with *p*MOS transistors results in a deeper current path and possibly a lower electric field, both reducing hot carrier effects. However, if constant-voltage scaling continues to apply down to submicron dimensions, the corresponding electric field becomes so high that hot carrier effect can begin to cause reliability problems even in a *p*MOS transistor. Recently, hot-electron induced punchthrough has been found important in submicron *p*MOS long-term reliability. Punchthrough voltage is significantly reduced due to hot electron injection into the gate oxide near the drain. This problem is serious

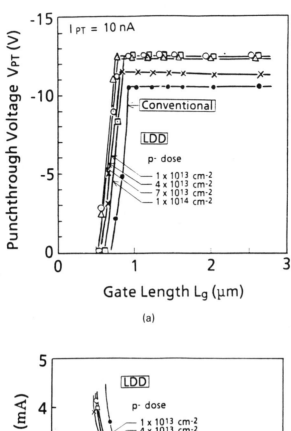

(a)

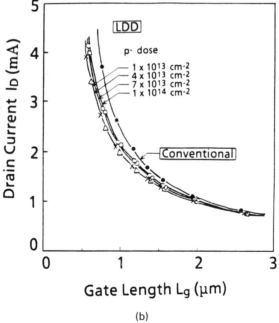

(b)

Figure 6.41 (a) Punchthrough voltage and (b) drain current as a function of gate length (L_g) for a conventional pMOS and LDD devices with various p^- doses (Ref. 58, © 1986 IEEE).

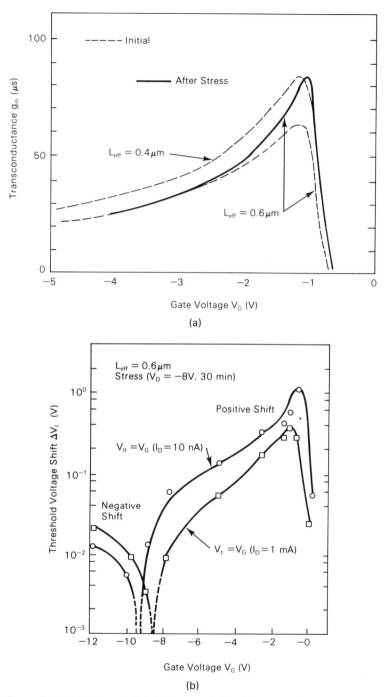

Figure 6.42 Submicron *p*MOS device degradation: (a) transconductance as a function of gate voltage before and after stress, stress condition: $V_{ds} = -8$ V and $V_G = -1.25$ V and (b) threshold shift as a function of stressing gate voltage (Ref. 59, © 1987 IEEE).

because, as previously described, high punchthrough voltage is hard to obtain for p-channel MOSFETs due to their buried channel nature. Additional punchthrough voltage reduction due to hot carrier injection further aggravates the problem.

The effect of hot carrier injection on p-channel device characteristics differs considerably from the hot carrier effects observed in n-channel devices.[61,62] In particular, the transconductance (g_m) of a p-channel device increases as a result of stress, and the magnitude of the threshold voltage decreases except at very high $|V_g|$.[59] Figure 6.42(a) and (b) show the stress-induced changes in g_m and V_t. At low $|V_g|$, the g_m of a 0.4 μm device is similar to that of a stressed device with 0.6 μm L_{eff}. This similarity is in contrast with the decrease in transconductance and increase in threshold voltage widely reported for n-channel devices. In an n-channel transistor, both channel hot electrons and hot holes generated by impact ionization are important in hot carrier induced device reliability.[26,63] However, in a p-channel device, hot holes do not play a significant role unless a device is stressed at large magnitudes of V_g, i.e., $|V_g| > |V_d|$.

As illustrated in Fig. 6.42(b),[58] threshold voltage shifts to a negative direction only at $V_g < -10$ V. The negative shift has also been observed for devices tested under accelerated stress for long periods of time.[64] In general, electrons are primarily responsible for the threshold shift because they are more likely to inject into the gate oxide than are holes due to their lower associated barrier heights, and at low V_g ($|V_g| < |V_d|$), the oxide field near the drain aids electron injection but retards holes.

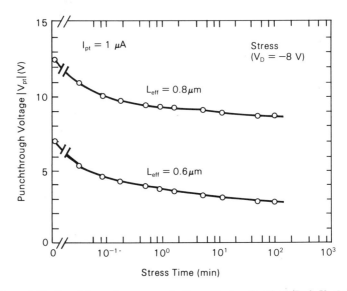

Figure 6.43 Punchthrough voltage reduction with stressing time (Ref. 59, © 1987 IEEE).

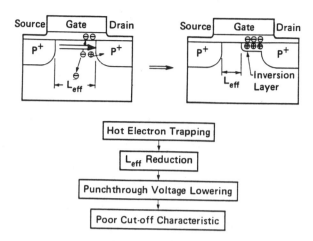

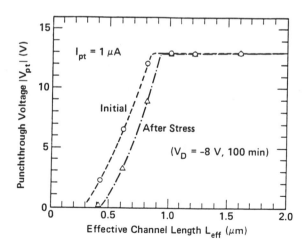

Figure 6.44   Schematic diagram illustrating hot-electron injection in a *p*MOSFET (Ref. 59, © 1987 IEEE).

In general, gate current measured in *p*MOS devices reaches a peak at low gate bias.   The worst stressing condition is at low $|V_g|$ (roughly -1.25 V) because its corresponding oxide field results in a maximum gate current and a maximum stress-induced V_t shift, although the maximum substrate current is at higher $|V_g|$ (roughly -3 V).   Again, this condition is opposite the results reported for *n*MOSFETs, in which device degradation is the worst at the gate voltage corresponding to the largest substrate current.

Another observation is the degradation of V_{PT} with stress time as shown in Fig. 6.43.   For devices with $L_{eff} = 0.6$ μm, punchthrough voltage rapidly

Figure 6.45   Punchthrough voltage as a function of initial effective channel length for *p*MOS devices before and after stressing (Ref. 59, © 1987 IEEE).

decreases to less than the supply voltage, 5 V. These observations can be explained by the schematic in Fig. 6.44. Hot electrons generated by impact ionization near the drain are accelerated toward the gate by the aiding oxide field and some of them become trapped in the oxide near the drain. The trapped electrons invert the Si surface near the drain resulting in an extension of the p^+ drain region. Consequently, effective channel length decreases, resulting in an increase in transconductance, but a degradation in punch-through and subthreshold characteristics.

Punchthrough voltages (V_{PT}) versus L_{eff} before and after stressing are shown in Fig. 6.45. Such punchthrough degradation is another outcome of the L_{eff} reduction caused by hot electron injection. The degradation corresponds to approximately 0.15 μm L_{eff} reduction, which is consistent with the approximately 0.2 μm reduction observed from the g_m measurement shown in Fig. 6.42(a). If the threshold voltage (V_t) is defined as V_g, which gives rise to 10 nA or 1 mA at a $V_{ds} = -5$ V, it is then sensitive to stress-

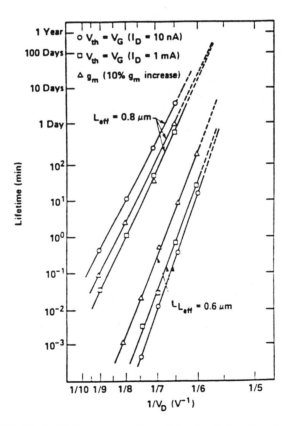

Figure 6.46 Device lifetime versus reciprocal stressing drain voltage for submicron *p*MOSFETs (Ref. 59, © 1987 IEEE).

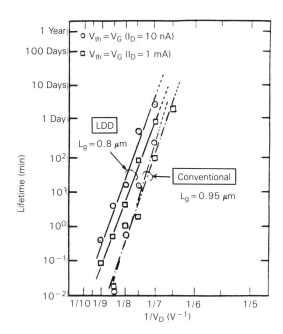

Figure 6.47 Device lifetime versus reciprocal stressing drain voltage for submicron LDD *p*MOS devices.   Data for a conventional *p*MOS device is also plotted for comparison (Ref. 59, © 1987 IEEE).

induced punchthrough current as well as oxide charges, and so is appropriate for worst-case evaluation.   The lifetimes, defined at 100 mV V_t shift, are shown for submicron *p*MOS devices (Fig. 6.46).   Lifetimes defined by 10% g_m increase are also plotted on the figure.[59]   In this example, lifetime is about one year for a 0.8 μm *p*MOSFET operating at the worst supply voltage (5 V + 10% variation).

To increase submicron *p*MOS lifetime, LDD structure is also applied. The use of LDD basically increases L_{eff}, hence compensating for L_{eff} reduction induced by hot-electron injection.   With a lightly-doped drain, it also reduces peak electric field.   As a result, punchthrough voltage is considerably improved when LDD is used.   Compared with conventional *p*MOSFETs, stress-induced V_t shifts in an LDD device are much smaller.[58]   The example shown in Fig. 6.47 implies that a lifetime of more than 10 years can be achieved in submicron *p*MOSFETs with an LDD structure.

REFERENCES

1. R. H. Dennard, F. H. Gaensslen, H. N. Yu, V. L. Rideout, E. Bassous, and A. LeBlanc, "Design of ion implanted MOSFETs with very small physical dimension," *IEEE J. Solid-State Circuits*, vol. SC-9, p. 256, 1974.

2. P. K. Chatterjee, W. R. Hunter, T. C. Holloway, and Y. T. Lin, "The impact of scaling laws on the choice of *n*-channel or *p*-channel for MOS VLSI," *IEEE Electron Device Lett.*, vol. EDL-1, p. 220, 1980.

3. M. Lenzlinger and E. H. Snow, "Fowler-Nordheim tunneling into thermally grown SiO_2," *J. of Applied Physics*, vol. 40, p. 278, 1969.

4. R. R. Troutman, "VLSI limitations from drain-induced barrier lowering," *IEEE Trans. on Electron Devices*, vol. ED-26, p. 461, 1979.

5. J. R. Brews, W. Fichtner, E. H. Nicollian, and S. M. Sze, "Generalized guide for MOSFETs miniaturation," *IEEE Electron Device Lett.*, vol. EDL-1, p. 2, 1980.

6. H. Shichijo, "A re-examination of practical performance limits of scaled *n*-channel and *p*-channel MOS devices for VLSI," *Solid State Electronics*, vol. 26, p. 969, 1983.

7. C. Y. Chang, Y. K. Fang, and S. M. Sze, "Specific contact resistance of metal-semiconductor barrier," *Solid-State Electron.*, vol. 13, p. 238, 1970.

8. J. Y. Chen, and D. B. Rensch, "The use of refractory metal and electron-beam sintering to reduce contact resistance for VLSI," *IEEE Trans. Electron Devices*, vol. ED-30, p. 1542, 1983.

9. H. H. Berger, "Models for contacts to planar devices," *Solid-State Electron.*, vol. 15, p. 145, 1972.

10. B. J. Sheu and P. K. Ko, "An analytical model for intrinsic capacitances of short-channel MOSFETs," in *IEDM Dig.*, p. 300, 1984.

11. J. D. Meindl, "Interconnection limits on silicon ultra large scale integration," *VLSI Sym. ECS*, Sp. 1986.

12. M. H. Woods, "Reliability in MOS integrated circuits," *IEDM Digest*, p. 50, 1984.

13. R. H. Dennard, F. H. Gaensslen, E. J. Walker, and P. W. Cook, "1 μm MOSFET VLSI technology: Part II—Device design and characteristics for high-performance logic applications," *IEEE Trans. on Electron Devices*, vol. ED-26, p. 325, 1979.

14. W. R. Hunter, L. Ephrath, W. D. Grobman, C. M. Osburn, B. L. Crowder, A. Cramer, and H. E. Luhn, "1 μm MOSFET VLSI technology: Part V—A single level polysilicon technology using electron beam lithography," *IEEE Trans. Electron Devices*, vol. ED-26, p. 353, 1979.

15. J. Y. Chen, R. C. Henderson, and D. E. Snyder, "A novel self-aligned isolation process for VLSI," *IEEE Trans. on Electron Devices*, vol. ED-30, p. 1521, 1983.

16. J. Y. Chen, "An *n*-well CMOS with self-aligned channel stops," *IEDM Digest*, p. 526, 1983.

17. T. Shibata, K. Hieda, M. Sato, M. Konake, R. L. M. Dang, and H. Iizuka, "An optimally designed process for submicrometer MOSFETs," *IEEE Trans. Electron Devices*, vol. ED-29, p. 531, 1982.

18. J. Zhu, R. A. Martin, and J. Y. Chen, "Subthreshold conduction in submicrometer buried channel P-MOSFETs," *IEEE Trans. Electron Devices*, vol. ED-35, p. 145, 1988.

19. T. Y. Chan, P. K. Ko and C. Hu, "Dependence of channel electric field on device scaling," *IEEE Electron Device Lett.*, EDL-6, p. 551, 1985.

20. P. K. Ko, R. S. Muller, and C. Hu, "A unified model for hot-electron currents in MOSFETs," in *IEDM Tech. Dig.*, p. 600, 1981.

21. C. Hu, S. C. Tam, F-C. Hsu, P. K. Ko, T. Y. Chan and K. W. Terril, "Hot-electron-induced MOSFET degradation—model, monitor and improvement," *IEEE J. of Solid-State Circuits*, vol. SC-20, p. 295, 1985.

22. P. K. Chatterjee, "VLSI dynamic NMOS design constraints due to drain induced primary and secondary impact ionization," in *IEDM Tech. Dig.*, p. 14, 1979.

23. W. Shockley, "Problems related to *p-n* junctions in silicon," *Solid-State Electron.*, vol. 2, pp. 35–67, 1961.

24. S. Tam, P. K. Ko, and C. Hu, "Lucky-electron model of channel hot electron injection in MOSFETs," pp. 1116–1125, *IEEE Trans. Electron Devices*, Sept. 1984.

25. T. H. Ning, P. W. Cook, R. H. Dennard, C. M. Osburn, S. E. Schuster, and H. N. Yu, "1 μm MOSFET VLSI technology: Part IV—Hot electron design constraints," *IEEE Trans. Electron Devices*, vol. ED-26, 346, 1979.

26. K. R. Hofmann, W. Weber, C. Werner and G. Dorda, "Hot carrier degradation mechanism in *N*-MOSFETs," in *IEDM Digest*, p. 104, 1984.

27. K. R. Hofmann, W. Weber, C. Werner and G. Dorda, "Hot-electron and hole-emission effects in short *n*-channel MOSFETs," *IEEE Trans. Electron Devices*, vol. ED-32, p. 691, 1985.

28. E. Takeda, H. Kume, T. Toyabe, and S. Asai, "Submicrometer MOSFET structure for minimizing hot-carrier generation," *IEEE Trans. Electron Device*, vol. ED-29, p. 611, 1982.

29. F. H. Hsu and K.-Y. Chiu, "A comparative study of tunneling, substrate hot-electron and channel hot-electron injection induced degradation in thin-gate MOSFETs," in *IEDM*, p.96, 1984.

30. E. Takeda and N. Suzuki, "An empirical model for device degradation due to hot-carrier injection," *IEEE, Electron Devices Lett.*, vol. EDL-4, p. 111, 1983.

31. F.-C. Hsu and K.Y. Chiu, "Hot-electron substrate-current generation during switching transient," *IEEE Trans. Electron. Devices*, vol. ED-32, p. 375, 1985.

32. T. K. Horiuchi, H. Mikoshiba, K. Nakamura, and Hamano, "A simple method to evaluate device lifetime due to hot-carrier effect under dynamic stress," *IEEE Trans. Electron Device Lett.*, vol. EDL-7, p. 337, 1986.

33. W. Weber, C. Werner and G. Dorda, "Degradation of *n*-MOS Transistors after pulsed stress," *IEEE Electron Device Lett.*, vol. EDL-5, p. 518, 1984; also, W. Weber, C. Werner and A. V. Schwerin, "Lifetimes and substrate currents in static and dynamic hot-carrier degradation," in *IEDM*, p. 390, 1986.

34. M. M. Kuo, K. Seki, P. M. Lee, J. Y. Choi, P. K. Ko, and C. Hu, "Simulation of MOSFET lifetime under AC hot-electron stress," *IEEE Trans. on Electron Devices*, vol. ED-35, p. 1004, 1988.

35. M. Koyanagi, H. Kaneko, and S. Shinizu, "Optimum design of n^+-n^- double-diffused drain MOSFET to reduce hot-carrier emission," *IEEE Trans. on Electron Devices*, vol. ED-32, p. 562, 1985.

36. S. Ogura, P. J. Tsang, W. W. Walker, D. L. Critchlow, and J. F. Shepard, "Design and characteristics of the lightly doped drain-source (LDD) insulated gate field-

effect transistor," *IEEE Trans. Electron Devices*, vol. ED-27, p. 1359, 1980; also, P. J. Tsang, S. Ogura, W. W. Walker, J. F. Shepard, and D. L. Critchlow, "Fabrication of high-performance LDDFETs with sidewall-spacer technology," *IEEE Trans. Electron Devices*, vol. ED-29, p. 590, 1982.

37. T. Y Huang, M. Koyanagi, A. G. Lewis, R. A. Martin, and J. Y. Chen, "Eliminating spacer-induced degradations in LDD transistors," Proceedings of International Symposium on VLSI Technology, Systems, and Applications, 1987 May 13–15; Taipei. Taiwan p. 260: Publishing Services: ERSO, 1987. Also, T. Y. Huang, J. Y. Chen, I. W. Wu and R. H. Bruce, "Reliability and performance of 1-μm n-channel transistors, ICL Internal Report, Xerox Palo Alto Research Center, 1986.

38. F. C. Hsu and K.-Y. Chiu, "Evaluation of LDD MOSFET's based on hot electron induced degradation," *IEEE Electron Device Lett.*, vol. EDL-5, p. 162, 1984.

39. P. K. Ko, T. Y. Chan, A. T. Wu, and C. Hu, "The effects of weak gate-to-drain (source) overlap on MOSFET characteristics," in *IEDM Dig.*, p. 292, 1986.

40. K. Mayaram, J. Lee, T. Y. Chan, and C. Hu, "An analytic perspective of LDD MOSFETs," *Sym. on VLSI Technology*, p. 61, 1986.

41. H. Mikoshiba, T. Horiuchi, and K. Hamano, "Comparison of drain structures in n-channel MOSFETs," *IEEE Trans. Elect. Dev.*, ED-33, p. 140, 1986.

42. Y. Tsunashima, T. Wada, K. Yamada, T. Moriya, M. Nakamura, R. Dang, K. Taniguchi, M. Kashiwagi, and H. Tango, "Metal-coated lightly-doped drain (MLD) MOSFET's for sub-micron VLSI's," in the *Dig. of Sym. on VLSI Technology*, p. 114, 1985.

43. C. F. Codella and S. Ogura, "Halo doping effects in submicron DI-LDD device design," in *IEDM Digest*, p. 230, 1985.

44. M. Nakahara, Y. Hiruta, T. Noguchi, M. Yoshida, K. Maeguchi, and K. Kanzaki, "Relief of hot carrier constraint on submicron CMOS devices by use of a buried channel structure," in *IEDM Dig.*, p. 238, 1985.

45. F. C. Hsu and K. Y. Chiu, "Effects of device processing on hot-electron induced device degradation," in *VLSI Sym. Digest*, p. 108, 1985.

46. L. C. Parrillo, S. J. Hillenius, R. L. Field, E. L. Hu, W. Fichtner, and M. L. Chen, "A fine-line CMOS technology that uses P^+ polysilicon/silicide gates for NMOS and PMOS devices," in *IEDM Digest*, p. 418, 1984.

47. L. C. Parrillo, "Process techniques and device design considerations for fabricating CMOS VLSI circuits," in *ICCD Digest*, p. 8, 1984.

48. K. Yu, "N-well CMOS for DRAM, UCLA Extension Course on VLSI: CMOS Technology," *UCLA*, Los Angeles, Ca., 1984.

49. J. Y. Chen and L. B. Roth, "Refractory metal and metal silicides for VLSI devices," *Solid State Technology*, p. 145, 1984.

50. N. Kobayashi, S. Iwata, N. Yamamoto and T. Terada, "A novel tungsten gate technology for VLSI applications," in the *Dig. of Sym. on VLSI Technology*, p. 94, 1983.

51. S. J. Hillenius and W.T. Lynch, "Gate material work function considerations for 0.5 micron CMOS," in the *Proc. of IEEE Inter'l. Conf. on Computer Design (ICCD): VLSI in Computer*, p. 147, 1985.

52. G. J. Hu, C. Y. Ting, Y. Taur, and R. H. Dennard, "Design and fabrication of p-channel FET for 1-μm CMOS technology," in *IEDM Digest*, p. 710, 1982.

53. T. N. Nguyen and J. D. Plummer, "A comparison of buried channel and surface channel MOSFETs for VLSI," 1981, Device Research Conf., *IEEE Trans. on Electron Devices*, vol. ED-29, P. 1663, 1982.

54. S. Y. Chiang, K. M. Cham, and R. D. Rung, "Optimization of submicron p-channel FET structure," in *IEDM Digest*, p. 534, 1983; also K. M. Cham, S. Y. Chiang, "Device design for the submicrometer p-channel FET with n^+ polysilicon gate," *IEEE Trans. Electron Devices*, vol. ED-31, p. 964, 1984.

55. A. Schmitz and J. Y. Chen, "Design, modelling and fabrication of subhalf-micrometer CMOS transistors," *IEEE Trans. Electron Devices*, vol. ED-33, p. 148, 1986.

56. Y. Taur, G. H. Hu, R. H. Dennard, L. M. Terman, C. Y. Ting, and K. E. Petrillo, "A self-aligned 1-μm-channel CMOS technology with retrograde n-well and thin epitaxy, *J. of Solid-State Circuits*, vol. SC-20, p. 123, 1985.

57. R. A. Martin and J. Y. Chen, "Optimized retrograde n-well for one micron CMOS technology," in the *Proc. of Custom Integrated Circuit Conf.*, p. 199, 1985.

58. M. Koyanagi, J. Zhu, A.G. Lewis, R.A. Martin, T.Y. Huang and J. Y. Chen, "Investigation and reduction of hot electron induced punchthrough (HEIP) effect in submicron pMOSFETs," in *IEDM Digest*, p. 722, 1986.

59. M. Koyanagi, A. G. Lewis, J. Zhu, R. A. Martin, T. Y. Huang, and J. Y. Chen, "Hot Electron Induced Punchthrough (HEIP) effect in Submicron PMOSFETs," *IEEE Trans. on Electron Devices*," vol. ED-34, p. 839, 1987.

60. S. Odanaka, M. Fukumoto, F. Fuse, M. Sasago, T. Yabu, T. Ohzone, "A new half-micrometer p-channel MOSFET with efficient punchthrough stops," *IEEE Trans. Elec. Dev.*, vol. ED-33, p. 317, 1986.

61. K. K. Ng and G. W. Taylor, "Effects of hot-carrier trapping in n- and p-channel MOSFETs," *IEEE Electron Devices*, vol. ED-30, p. 871, 1983.

62. T. Tsuchiya and J. Frey, "Relationship between hot-electrons/holes and degradation for p- and n-channel MOSFETs," *IEEE Electron Device Lett.*, Vol. EDL-6, p. 8, 1985.

63. P. E. Cotrell, R. R. Troutman, and T. H. Ning, "Hot-electron emission in n-channel IGFETs," *IEEE Trans. Electron Devices*, vol. ED-26, p. 520, 1979.

64. J. J. Tzou, C. C. Yao, and H.W.K. Chan, "Hot-carrier-induced degradation in p-channel DD MOSFETs," *IEEE Electron Device Lett.*, vol. EDL-7, p. 5, 1986.

EXERCISES

1. Derive the change of delay time, power, power density, and power-delay product as a result of constant-voltage and quasi constant-voltage scaling of a MOSFET.

2. How does a MOSFET's threshold change when constant-field scaling rules are applied? Does it scale proportionally?

3. For a 0.5 μm n-channel MOSFET with $t_{ox} = 200$ Å, $W = 10$ μm, and $\mu_n = 500$ cm²/V-sec, if the source/drain sheet resistance is 100 $\Omega/\square$ and the source or drain

contact to poly-gate distance is 10 μm, calculate the intrinsic transistor channel conductance (g_d), transconductance(g_m), and total g_d and g_m in which source and drain resistances are included (assuming contact resistance is negligible).

4. Describe punchthrough current and subthreshold current and explain the differences between them.

5. An n-channel MOSFET has 200 Å t_{ox}, 1 μm L_{eff}, 0.2 μm x_j and $V_t = 0.6$ V. If the device is biased at $V_g = 3$ V and $V_d = 5$ V, calculate the saturation voltage V_{dsat} and maximum electric field E_m.

6. (a) Write the E_m expression for an LDD nMOSFET and using the E_m value calculated in Exercise 5, calculate the decrease of E_m if the n^- LDD length is 0.2 μm. (b) If a FET channel width is 10 μm and the S/D sheet resistance in the LDD region is 1000 $\Omega/\square$, calculate additional S/D resistance due to the LDD insertion and compare it with the intrinsic FET resistance.

7. Describe the process sequence for an LDD CMOS process and state qualitatively the advantages and disadvantages of an LDD structure.

8. If n^+ polysilicon is used for both n- and p-channel FET gates in a CMOS chip, what are the threshold voltages for the n- and p-channel devices assuming 250 Å gate oxide, 5×10^{16} cm^{-3} substrate doping concentration, and a fixed oxide charge density of 5×10^{10} cm^{-2}?

9. For Exercise 8, what are the threshold voltages if p^+ polysilicon is used for both n- and p-channel devices?

7

CMOS ISOLATION

The preceding chapter discussed device design for individual n- and p-channel transistors. Because as many as hundreds of thousands of transistors are used in a CMOS VLSI chip, they must be properly isolated so that the conduction state for each transistor can be individually controlled. What is unique in a CMOS circuit is the isolation among opposite-type transistors as well as among similar-type transistors. This chapter starts with the physics and techniques of field isolation in MOS technologies. It then discusses isolation in CMOS including physical and electrical isolation, failure mechanisms, dependence on process architecture, and new isolation techniques for future CMOS.

7.1 BACKGROUND

One important aspect of integrated circuit technology is the art of isolation among transistors which are all built in a common silicon substrate. The state (on or off) and conductance of each transistor can only be controlled if proper isolation exists among transistors. If not, leakage current paths occur, causing DC power dissipation, noise margin degradation, and voltage shift on a dynamic node. Crosstalk among transistors can destroy the logic state of each device. Dynamic RAMs are very sensitive to leakage because it can degrade memory holding time, hence requiring more refresh cycles. In a CMOS circuit, leakage current in the isolation region can escalate latchup, an important phenomenon to be described in the next chapter.

Device isolation is becoming more critical in VLSI because of increased transistor counts and decreased isolation space. For a VLSI chip, a small amount of leakage per transistor can induce significant power dissipation for the entire chip. For example, if a VLSI chip has one million transistors with each transistor producing 1 μA leakage, the total power dissipation for the entire chip would increase on the order of a few watts. Because the transistors are closely packed on a VLSI chip, adequate isolation between neighboring devices is more difficult.

In MOS technologies, isolation is commonly formed using a thick oxide and/or a heavily-doped Si layer in the field region among active MOSFETs (Fig. 7.1). The thick oxide is often called field oxide and the heavily-doped region is often referred to as channel stop. This field isolation works well for isolating active MOSFETs if only the Si surface in the active channel region is allowed to invert. This inversion occurs when the gate voltage reaches the threshold voltage of the active MOSFETs. As described in Chapter 2, threshold voltage increases with the increase of oxide thickness or substrate doping concentration according to Eq. 2.11. Thus, threshold voltage in a field region consisting of a thick oxide and/or heavily doped region is normally around 10 V or more, at least an order of magnitude higher than the threshold of active MOSFETs. As a result, surface inversion does not occur under the field oxide because the polysilicon or metal lines on the oxide is normally at 5 V or below. In normal operations, a conducting channel is allowed to form only in the active area, not in the field region.

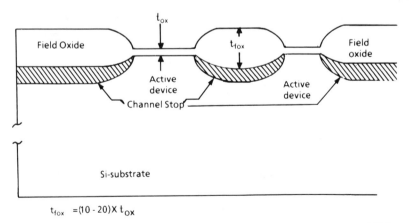

$$t_{fox} = (10-20) \times t_{ox}$$

Figure 7.1 Schematic of field isolation with self-aligned channel stop for MOS ICs.

7.1.1 **Test Structures**

This section considers NMOS technology only. Shown in Fig. 7.2 is an IC structure consisting of an aluminum interconnect line running between two active *n*-channel MOSFETs. Both a top-view layout and a cross-sectional view are shown. The aluminum line acts as a gate on the thick field oxide and the two n^+ diffusions adjacent to it act as source and drain, forming a parasitic FET. This FET is commonly used as a test structure to detect leakage current in the field region under the aluminum interconnect lines. Because polysilicon lines are also used for interconnects in an IC, a similar structure as shown in Fig. 7.3 is also possible. Notice the subtle difference as the n^+ regions adjacent to the polysilicon gate do not abut to the field oxide. This exact structure in general may not exist in an actual IC layout because polysilicon lines normally do not extend over the active (thin oxide) area (see Fig. 7.4(a)) unless they are also used for the gate electrode of active transistors (Fig. 7.4(b)). Here, the polysilicon line serves as gate electrode as well as the interconnect between the two electrodes. In Fig. 7.3, the

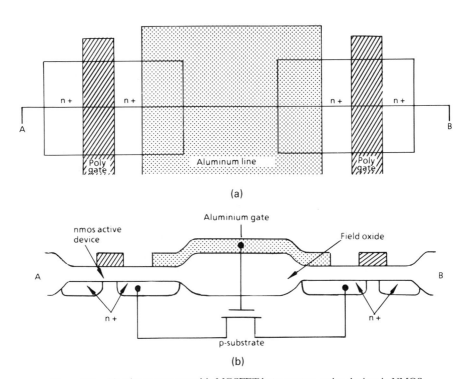

Figure 7.2 Aluminum-gate parasitic MOSFET between two active devices in NMOS technology: (a) layout; and (b) cross-sectional view.

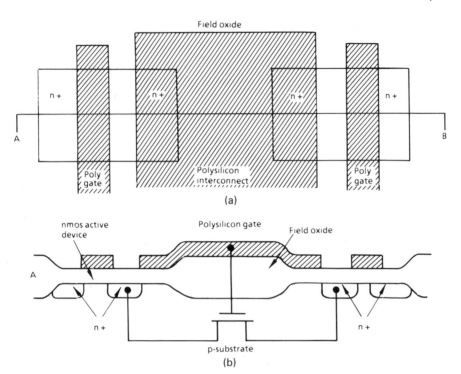

Figure 7.3 Polysilicon-gate parasitic MOSFET between two active devices in NMOS technology: (a) layout; and (b) cross-sectional view.

silicon surface under the thin oxide is normally inverted at a very low gate voltage (≤ 1 V), resulting in ultra shallow "n^+" regions. If the silicon surface under the field oxide is inverted, the two transistors are basically shortened and isolation between the two transistors is completely broken down. Thus, the test structure shown in Fig. 7.3 is a good test vehicle for the case shown in Fig. 7.4(b), because leakage current measured from the parasitic FET indicates the onset of surface inversion under the field oxide.

For the layout shown in Fig. 7.4(a), the test structure that has polysilicon overlapping the field oxide also tests the worst case because inversion is harder to achieve under field oxide that is not covered with a polysilicon line (see areas pinpointed by arrows in Fig. 7.4(a)). The configuration in Fig. 7.4(a), which is often seen in a real circuit, is unlikely to have an isolation problem if the test structure in Fig. 7.3 passes the isolation test.

Another difference between the aluminum and polysilicon gate parasitic FETs is the thicker oxide in the former devices due to an additional oxide layer deposited to separate polysilicon and aluminum interconnects. The total oxide under the aluminum is typically $1-2$ μm, while in the polysilicon device, the gate oxide is composed of thermally grown oxide only, which is

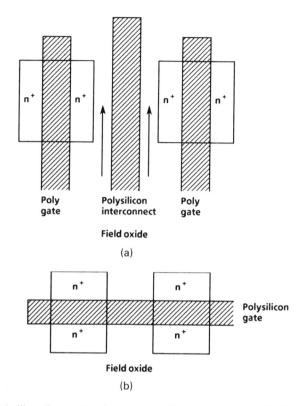

Poly
gate

Polysilicon
interconnect

Poly
gate

Field oxide

(a)

**Polysilicon
gate**

Field oxide

(b)

Figure 7.4   Polysilicon lines as transistor gate; and interconnect in real NMOS IC layouts.

about 0.5–1 μm.   In both cases, the thick field oxide and the doped silicon surface underneath inhabit surface inversion unless the gate voltage is raised several volts above the 5 V supply.

7.1.2 Field Threshold Voltage

Field oxide threshold voltage (V_{TF}), as cited in the literature to evaluate the rigorousness of field isolation, has often been defined as the voltage required to get 1-μA current in a parasitic FET.   This definition is crude and as such V_{TF} of >5 V does not necessarily correspond to an adequate isolation for circuits operated at 5 V power supply.   Actually, the subthreshold swing of the parasitic thick-oxide FET is so large (e.g., $S_t = 0.5$–0.6 V/DEC) that the circuit must operate at 3–4 V below V_{TF} to ensure a <1 pA off-state current.   In other words, V_{TF}, so defined, must be at least 8–9 V for a 5 V circuit operation.   This condition must be met for minimum isolation spacing in a given technology, because V_{TF} decreases with isolation spacing in a similar way to the short-channel effect observed in active FETs.[1] Even worse, as temperature increases, V_{TF} decreases and the subthreshold swing (V/DEC) rises.[2] For example, V_{TF} reductions as much as 2 V were noted as the temperature was increased from 25 to 125°C.[1]

Finally, another problem exists if circuits are used in a radiation environment, because the threshold shift, caused by radiation-induced oxide charges, is proportional to the oxide thickness to a power of between 2 and 3.[3] The resulting threshold shift for a field oxide can be 5 to 10 V depending on the total dose of radiation. It is therefore essential to provide an additional margin for radiation-hard circuits in which field threshold voltage as high as $\approx$ 20 V may be necessary. A more stringent measure of isolation will be discussed in Sec. 7.3.

7.2 MOS ISOLATION TECHNIQUES

An ideal isolation technique must satisfy all of the following requirements. First, leakage current must be negligible between active devices. For MOS technology, the silicon surface under the field oxide cannot be inverted during circuit operation. Second, spacing between active devices should be minimal. This requirement is particularly important for VLSI isolation. Third, isolation should not consume a significant portion of the active device area or result in significant narrow-channel effect as described in Chapter 2. In other words, the transition from an isolation area to the neighboring active area should be abrupt. Fourth, the isolation process should not adversely affect the process parameters required for fabricating the active device. In this way, the performance of an active device can be optimal. Finally, a process should not be too complicated to control. For some applications, the low leakage requirement should be met in high-temperature or radiation environments.

7.2.1 Window or Moat Isolation

This method involves etching windows for the active devices in a uniform thick oxide. This technique was used before LOCOS[4] was invented and was an industrial standard. The process is simple if a channel stop is not incorporated. Although sufficiently high field threshold voltage can be achieved with a thick enough field oxide alone, step coverage becomes a serious problem. Metal or polysilicon lines at the sharp edge of the oxide steps tend to break, especially when they are narrowed down for VLSI interconnects. Techniques such as implanting argon ions near the top portion of the oxide have been used to change the etching rate for the formation of a sloped oxide edge. This method facilitates step coverage, however, at the expense of a larger layout area to accommodate the gradual transition from the active to isolation region. Obviously, reduced oxide thickness is most effective to ease the step coverage problem, but then a channel stop must be incorporated to produce adequate field threshold voltage.

For p-channel MOS technology, a window isolation structure with self-aligned channel stop was developed.[5] In this structure, a phosphorus implant over the entire wafer is performed after the windows are etched into the field oxide. The phosphorus-doped active region is compensated by a shallow boron implant that goes only through windows patterned in the field oxide. The phosphorus implant forms an n-type channel stop under the field oxide, whereas the boron implant through the windows provides adequate channel doping for active devices. This process is similar to the retrograde n-well CMOS process as described in Chapter 5. Good device characteristics were obtained for MOSFETs with 8-μm channel length. However, the two implant conditions must be controlled precisely to achieve the exact doping compensation for threshold control. A 2 σ variation of 0.35 V was found in earlier work.[5] This control problem becomes more serious for scaled devices, which require higher implant doses and lower threshold voltage.

Similar window isolation has been done for NMOS technology. Active channel regions are first defined by etching windows in a thick field oxide. Boron ions are deeply implanted under the oxide to form a p-type channel stop.[6] The boron impurities in the active regions are counter doped by a shallow phosphorus implant to set a low threshold voltage. Again, this process is like a retrograde p-well process. The cross-sectional view of the final device structure is shown in Fig. 7.5. Although submicron MOSFETs with adequate threshold and punchthrough voltages were made, they had poor body effect and high junction capacitance because of excessive boron impurities in the channel region. Moreover, a significant 2-D fringing field effect existed in the devices with narrow channel widths. Figure 7.6 shows the 2-D contours for potentials and carrier concentrations in a narrow-channel device with its gate biased slightly above threshold voltage. Only half of the channel is shown due to the symmetry of the structure. The cross-section is perpendicular to the direction of current. Even with a precisely-defined field oxide edge, both surface potential and electron concentration decrease significantly near the edge of the channel, resulting in a reduction of effective (or electrical) channel width. This channel-narrowing effect is the origin of threshold increase and transconductance decrease as the channel width is reduced.

In spite of the problems previously mentioned, direct window isolation provides advantages for isolation in high-density ICs. Compared to LOCOS, another isolation technique, direct window (moat) isolation has been shown to save 30%–40% of the silicon area.[1] It was pointed out that increasing doping concentration in short-channel ($\approx$1 μm) MOSFETs relaxed the requirement of N_A (channel stop) $> N_A$ (channel), where N_A represented the

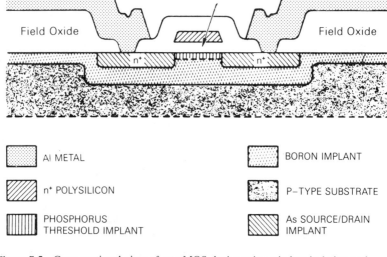

AI METAL

n⁺ POLYSILICON

PHOSPHORUS
THRESHOLD IMPLANT

BORON IMPLANT

P–TYPE SUBSTRATE

As SOURCE/DRAIN
IMPLANT

Figure 7.5 Cross-sectional view of an *n*MOS device using window isolation and channel stop.

dopant concentration.[1,7] It was therefore possible, using a single boron implant, to form a channel stop under the field oxide and provide punchthrough protection in the channel region.

Another implant was made to adjust threshold voltage after the active areas were defined. This technique, however, was not completely satisfactory for an optimum NMOS process. Neither channel stop nor channel area can have optimized doping distribution by using only one implant. The boron implants selected in the work[1] may have offered suitable punchthrough protection, but the boron profile was too deep to provide appropriate body effect and junction capacitance for the active devices. Also, the dopant concentration for the channel stop might not have been sufficient.

Another NMOS process that forms channel stops for device isolation, but independently dopes channel regions for active devices, has been proposed.[8] After etching windows into field oxide with a patterned photoresist layer, a layer of metal is deposited and the portion above the field oxide is subsequently lifted off, leaving metal just inside the windows. The metal inside the windows acts as an implant mask to block boron ions while they are implanted through the field oxide to form channel stops. The channel stops are self-aligned to the thick field oxide for high packing density. Meanwhile, the active devices can have an optimum impurity profile. The associated liftoff process is, however difficult for VLSI production.

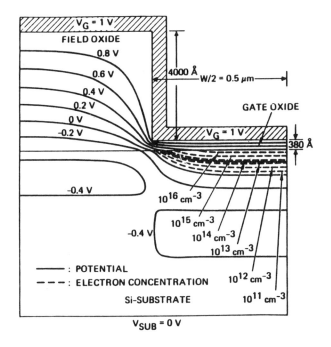

Figure 7.6 Two-dimensional contours of equipotentials and electron concentrations for a non-planar MOS structure in channel width direction (Ref. 13, © 1982 IEEE).

7.2.2 LOCal Oxidation of Silicon (LOCOS)

The LOCOS (LOCal Oxidation of Silicon) process[4] is commonly used for the isolation of active devices in the MOS IC industry.   As shown in Fig. 7.7, this process uses a patterned nitride layer together with a thin buffer oxide underneath to mask the active area.   This pad oxide releases stress caused by the nitride layer, and is therefore often referred to as SRO for Stress-Relief Oxide.   The nitride/oxide mask is used to block ions during channel stop implantation and the same mask is then used for local oxidation. The nitride/oxide masking layer is stripped after LOCOS.   The resulting isolation structure consists of a semi-recessed field oxide with the channel stop naturally self-aligned to the oxide edges.   This standard LOCOS process can be modified to produce fully recessed field oxide if the Si substrate in the isolation region is etched before LOCOS.   Normally, the amount of Si etched is roughly half of the desired field oxide thickness so that an isoplanar surface is formed after LOCOS.   This scheme however adds more processing steps and concerns for etch-induced crystal damage and increased oxidation encroachment.

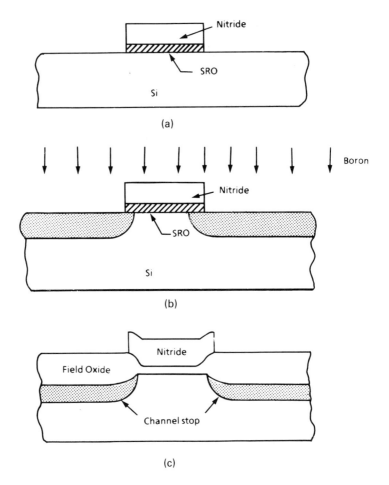

Figure 7.7   Process steps for LOCOS isolation with self-aligned channel stop.

Applying LOCOS to VLSI isolation is severely limited by field oxide encroachment and lateral diffusion of channel-stop dopants into the active device area.   Oxidant diffuses through the buffer oxide under the nitride, hence inducing oxidation near the nitride edges.   This lateral oxidation encroachment makes the edge of the LOCOS oxide resemble a bird's beak. The impurity atoms in the channel stop region also diffuse laterally during LOCOS and other subsequent high temperature process steps.   Figure 7.8 is an SEM micrograph showing the bird's beak structure due to lateral oxidation.   As shown in Fig. 7.9, the lateral oxidation rate is a function of the lateral distance from the nitride edge and depends on buffer oxide thickness. Both oxidation encroachment and impurity lateral diffusion overtake the active area near the channel edges.   Hence, the physical channel width W_p

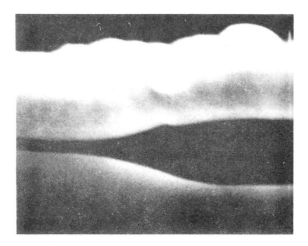

Figure 7.8 SEM micrograph showing the cross-section of the bird's beak due to oxide encroachment (Ref. 9, © 1981 IEEE).

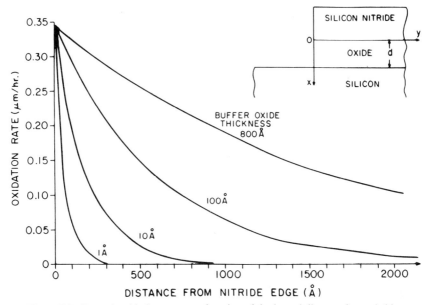

Figure 7.9 Lateral oxidation rate as a function of the lateral distance from nitride edge for various buffered oxide thicknesses used in the LOCOS process (Ref. 9, © 1981 IEEE).

becomes less than the designed channel width W_d. The difference (W_d – W_p) becomes a significant portion of W_d when devices are scaled down for VLSI applications. Furthermore, the electrical channel width (W_e) is even less than W_p because the 2-D narrow-channel effect applies to LOCOS as well as other non-planar isolation schemes, such as window isolation. Both physical and electrical channel-narrowing effects increase the FET threshold voltage and reduce its current driving capability. To maintain the desired channel width, the corresponding mask dimension must be drawn oversized, resulting in wasted layout area. Other considerations are nitride-induced crystal defects and step coverage on non-planar surface.

Several methods have been tried to improve LOCOS isolation in terms of reducing the bird's beak. Sealed-Interface Local Oxidation (SILO)[9] uses a thin (<300 Å) nitride film directly on Si to seal the interface, shutting off lateral oxidant diffusion during local oxidation. As predicted in Fig. 7.9, the lateral oxidation rate decreases sharply as the buffer oxide thickness is reduced. Consequently, a steeper step angle and reduced bird's beak[9] can be obtained (Fig. 7.10). A thicker nitride layer can reduce the bird's beak even further, but at the expense of generating more stress, more crystal defects, and higher leakage. To achieve low stress and a small bird's beak, a sandwich masking layer (Fig. 7.11) consisting of a thin nitride (100–200 Å), an LPCVD (Low-Pressure Chemical Vapor Deposited) oxide (300 Å) and a thicker nitride (1000 Å) has shown basically defect-free field oxide with encroachment of only 25% of the oxide thickness (Fig. 7.12). A conventional LOCOS generally has lateral encroachment of ≈80–100%.

Another attempt to improve LOCOS is the use of the SideWAll Masked Isolation (SWAMI) process,[10] which offers basically defect-free and near-zero bird's beak, but at the expense of process complexity. As shown in Fig. 7.13, the process starts with forming and patterning a composite masking layer with stress relief oxide (SRO I), and nitride (Nitride I) on active regions. The Si substrate is then etched with a sloped sidewall and channel stop is implanted (Fig. 7.13a). A second composite nitride/oxide layer is then formed

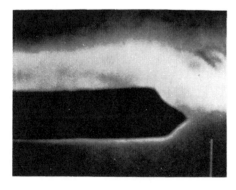

Figure 7.10 An SEM micrograph showing the cross-section of reduced bird's beak in a SILO structure (Ref. 9, © 1981 IEEE).

SILO

Figure 7.11 Process steps for a sand-wiched SILO structure.

and an LPCVD oxide is deposited (Fig. 7.13b).   After RIE etching of the CVD oxide, an oxide sidewall spacer is left (Fig. 7.13c). This sidewall spacer is then used as a mask to etch the nitride II, then the mask is stripped (Fig. 7.13d). LOCOS is then performed with the Si sidewall protected from lateral oxidation encroachment by the thin nitride spacer (Fig. 7.13e). All masking materials are then stripped and gate oxide is grown (Fig. 7.13f). The process employs two nitride masks, anisotropic oxide and nitride etching. The slope of the sidewall ($\approx 54°$) must be controlled during Si etching. A near isoplanar structure is achieved as shown in Fig. 7.14 and good electrical characteristics have been obtained.[10]

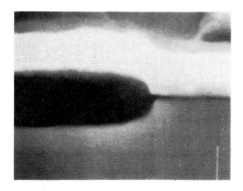

Figure 7.12 SEM micrograph showing the cross-section of minimum encroach-ment using the sandwich structure shown in Fig. 7.11 (Ref. 9, © 1981, IEEE).

THE SWAMI

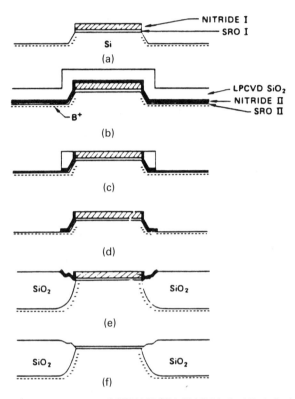

Figure 7.13 The major process steps of SWAMI (SideWAll Masked Isolation) (Ref. 10, © 1982 IEEE).

7.2.3 Isoplanar Field Oxide

Another isolation approach involves etching a shallow (several tenths of a micron) step in a silicon substrate and filling it with a depositing oxide. This isolation has fully recessed isoplanar configuration and is bird's beak-free. It was introduced as BOX for Buried OXide.[11] A liftoff technique using aluminum[11] or molybdenum[12] as a stencil was employed for this approach. As shown in Fig. 7.15, SiO_2 was plasma-deposited at 300°C for the field oxide. A wet etching before liftoff was done to remove oxide at the sidewall of the Si groove. Because the oxide at the sidewall was highly stressed, its etch rate was much higher, allowing preferential etching. The oxide on the metal stencil was then lifted off by resolving the metal. To fill

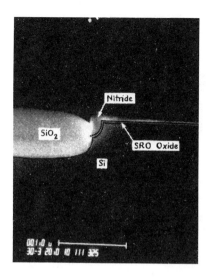

Figure 7.14 SEM micrograph showing the cross-section of a SWAMI structure (Ref. 10, © 1982 IEEE).

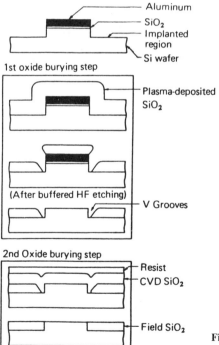

Figure 7.15 Process steps in BOX (Buried OXide) technology (Ref. 11, © 1981 IEEE).

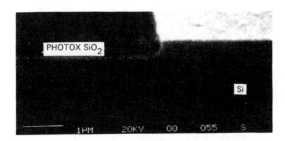

Figure 7.16 SEM micrograph showing the cross-section of isoplanar PHOTOX structure (Ref. 13, © 1982 IEEE).

in the V-grooves at the step, another oxide was deposited and RIE etched to form a planar surface.

A similar technique using photoresist to lift off low-temperature (100°C) photo-CVD oxide (Photox) was also proposed for field isolation.[13] The idea again is to provide a fully isoplanar structure as shown in Fig. 7.16. The

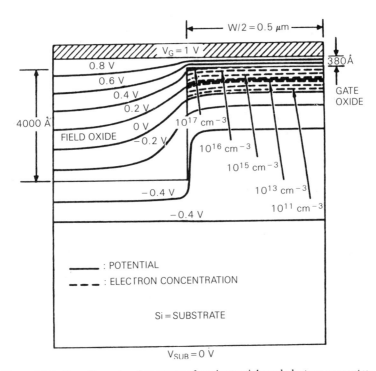

Figure 7.17 Two-dimensional contours of equipotentials and electron concentrations for a fully-recessed isoplanar MOS structure in channel width direction (Ref. 13, © 1982 IEEE).

planar isolation with fully-recessed field oxide not only eliminates physical encroachment and surface topology, but also avoids the electrical 2-D fringing field effect illustrated in Fig. 7.6.   Fig. 7.17 shows the 2-D contours of potential lines and carrier concentrations for the planar isolation structure.   Notice that both potentials and electron concentrations are uniform across the entire channel width, indicating negligible electrical channel-narrowing effect.

Despite all of the advantages of the isoplanar structure, the associated liftoff process is, so far, doubtful for production.   The use of preferential etching can facilitate liftoff, but even then the process may be too complicated for manufacturing VLSI circuits.   Other processes, such as trench, are being developed for realizing isoplanar isolation.   Section 7.4 will discuss these techniques.

Table 7.1 summarizes the various MOS isolation techniques discussed. Key process steps, distinct features, and schematics are included.   For the present MOS IC fabrication, standard LOCOS or improved LOCOS such as SILO, is still the dominant isolation technique, even though scaling isolation spacing for VLSI becomes severe.

7.3 ISOLATION IN CMOS

Layout density has been a major problem in bulk CMOS because both n- and p-channel devices must be made on the same chip and the two devices must be properly isolated.   The two types of devices must be separated physically by at least one well, which is normally formed by diffusion.   Because diffusion is an isotropic process, a large layout space is required to accommodate lateral diffusion during well formation.   Another problem is the implementation of channel stop(s) for CMOS devices.   In NMOS technology, the p-type channel stop can be formed easily under the LOCOS oxide to raise field threshold voltage well above the 5 V power supply.   In the case of CMOS, a p-type channel stop is still necessary to provide isolation among n-channel transistors.   Moreover, it must be spatially restricted to the n-channel area only.   Similar requirements apply for n-type channel stops needed in the p-channel area.   These additional constraints require extra layout spacing between the n- and p-channel areas unless some sort of self-aligned structure can be realized.   Even worse, the channel-stop doping concentration at the edge of the well drops substantially due to lateral diffusion and the compensation between acceptor and donor impurity atoms.   As an example, Fig. 7.18 shows this effect through 2-D simulation for the twin-tub process.[14] The solid curves show the reduction in net impurity concentration at the well boundary when both boron and phosphorus implants are used.   The dashed line represents a case in which only boron implant was used.   The doping reduction leads to isolation problems associated with the tight spacing between an n- and p-channel device.[14]

TABLE 7.1 Comparison of Various MOS Field Isolation Technologies

TECHNIQUE		KEY PROCESS STEPS	DISTINCTIVE FEATURES
LOCOS	NITRIDE ON OXIDE		BIRD'S BEAK SEMI-RECESSED
IMPROVED LOCOS	ELIMINATING BUFFER OXIDE		• REDUCED BIRD'S BEAK LENGTH • DIFFICULT TO CONTROL DEFECT FORMATION • SURFACE NOT COMPLETELY FLAT (PLANAR)
	USING NITRIDE PROTECTION OF SIDEWALL		• EXCELLENT SURFACE PLANARITY • VIRTUALLY NO BIRD'S BEAK • DEFECT FORMATION (?) • COMPLICATED
WINDOW ISOLATION			• VIRTUALLY NO BIRD'S BEAK • SIMPLICITY OF PROCESS • SURFACE NOT FLAT (PLANAR) • DIFFICULT TO APPLY TO SUBMICRON ISOLATION • STEP COVERAGE
GROOVE	BURIED OXIDE		• NO BIRD'S BEAK • PERFECTLY FLAT (PLANAR) SURFACE • NO IMPURITY REDISTRIBUTION • REQUIRE Al MASK AND PLANARIZATION BY RIE
	PHOTOX		• NO BIRD'S BEAK • SUPER-PLANAR • NO Al OR RIE • NOT MATURE
TRENCH			• EFFECTIVE FOR BIPOLAR OR CMOS WELL ISOLATION • INCREASED PARASITIC INTERCONNECT CAPACITANCE • COMPLICATED • $\geq 2\,\mu m$ WIDE?

Si₃N₄: Si_3N_4 POLY-Si   SiO₂: SiO_2 Si

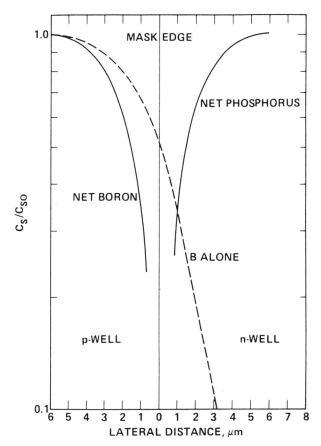

Figure 7.18 Impurity surface concentration near the border of the two tubs in twin-tub CMOS (Ref. 14, © 1980 IEEE).

In a CMOS circuit, proper isolation must be provided among opposite-type devices as well as between similar devices.[15] The following sections first describe isolation among similar devices, followed by the discussion of isolation among opposite-type devices, then the enhancement of isolation by using retrograde-well technologies and finally, vertical isolation between diffusion in the well and the substrate.

7.3.1 Isolation Among Similar Transistors

This section will discuss isolation between two *n*-channel transistors and two *p*-channel transistors. These similar transistors must be isolated properly to avoid leakage current occurring in parasitic FETs. Parasitic FETs are formed by polysilicon or aluminum interconnect lines on relatively thick field oxide and adjacent *pn* junctions.

Test devices for isolation between two *n*-channel active transistors were described in Sec. 7.1.1. Problems associated with the conventional definition of field threshold voltage were discussed in Sec. 7.1.2. Instead of defining threshold at 1 μA, a stringent way to test field isolation is to measure the field threshold voltage (V_{TF}), defined as the gate voltage that generates a small and acceptable drain current at drain voltage equal to power supply voltage, i.e., currently a 5 V V_{DD}. The V_{TF} values measured at drain currents of 10 pA, 1 nA and 0.1 μA in a 20 μm wide FET are shown in Fig. 7.19.[16] The test structures used for this measurement are shown in Figs. 7.2 and 7.3. The LOCOS field oxide thicknesses for polysilicon and aluminum gate FETs are about 0.5 and 1.0 μm, respectively in this example. A boron field implant of 4×10^{13} cm^{-2} at 25 KeV is used for the channel stop formation before the LOCOS oxidation. The higher current (0.1–1.0 μA) corresponds to a threshold similar to the threshold defined at the onset of strong inversion. This definition is suitable for the active MOS transistors but not particularly useful for parasitic FETs in which a very low level of leakage current (10 pA – 1 nA) is a serious concern for maintaining isolation. Notice that the V_{TF}s are higher for the aluminum gate because of the associated thicker field oxide. The thicker oxide also gives poorer subthreshold slope, causing V_{TF} curves to separate further. The V_{TF} decreases as the channel length (n^+-to-n^+ separation) is reduced in a way that is analogous to the short-channel effect commonly observed in active MOSFETs (see Chapter 2, Sec. 2.5). This effect should be taken into account when setting n^+-to-n^+ layout design rules. For the example shown in Fig. 7.19, V_{TF} values are 10 V or above even at 1 μm separation between two *n*-channel active devices.

Parasitic FETs among *p*-channel active devices are identical to those shown in Figs. 7.2 and 7.3 except that the doping types are reversed. The symmetric parasitic pMOSFETs are, in general, harder to turn on because the net oxide charge density (Q_{fc} term) is commonly positive, making the V_{TF} even more negative. As shown in Fig. 7.20,[16] the absolute V_{TF} values are well above 5 V. The test devices used here are fabricated from the same CMOS process which made the devices used for obtaining data in Fig. 7.19. The channel stop is formed by a shallow phosphorus implant inside the *n*-well. The very high V_{TF} values in the polysilicon gate devices are not always measurable due to dielectric breakdown in the thin oxide region between the polysilicon gate and the source. Notice that the aluminum gate devices exhibit greater short-channel effect, due to thicker oxide.

Leakage can also be observed in the parasitic FETs if punchthrough between the two n^+ (or two p^+) regions occurs. Due to the relatively high doping concentration in the field region and shallow n^+ or p^+ diffusion regions, punchthrough usually does not take place unless the separation between the two diffusion regions is scaled below 1 μm. Breakdown is normally

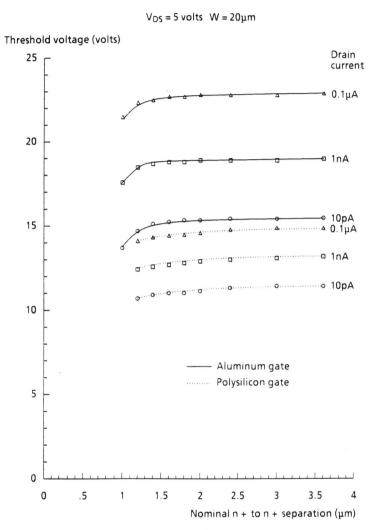

Figure 7.19 n-Channel field threshold voltages for symmetrical parasitic FETs (Ref. 16).

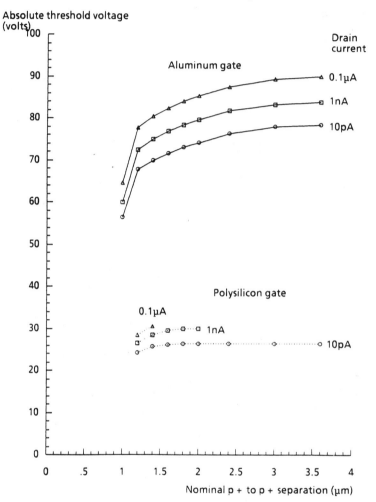

V_{DS} = -5 volts   W = 20µm

Figure 7.20 *p*-Channel field threshold voltages for symmetrical parasitic FETs (Ref. 16).

limited by avalanche breakdown which, in general, occurs at voltages well above 5 V.

7.3.2 Isolation Among Opposite-Type FETs

In addition to physical separation and the precise definition of channel stops, electrical isolation between n-channel and p-channel active devices is just as important.   Inadequate isolation leads to leakage between opposite-type devices.   This leakage corresponds to the subthreshold leakage measured in parasitic FETs.

Test structures.   Fig. 7.21 shows two parasitic FETs formed at the edge of an n-well by aluminum or polysilicon lines.   The n^+ and the n-well form source and drain for the parasitic nMOSFET while the p^+ and p-substrate serve as source and drain for the pMOSFET.   Here, the source and drain are not symmetrical and the n-well and p-substrate are used as drains because a deeper drain junction and a highly conductive source always produce more subthreshold current, thereby representing a worst case condition for isolation.   Similar to the test structures described in Sec. 7.3.1, the two structures have two differences between them.   The field oxide in the aluminum gate structure is thicker than that in the polysilicon gate.   Also, the

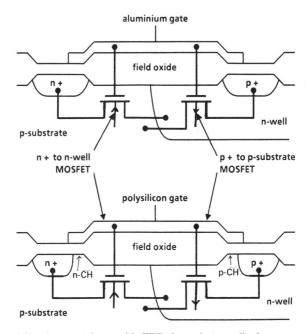

Figure 7.21 Asymmetric parasitic FETs formed at n-well edge.

n^+ and p^+ in the polysilicon gate structure are self-aligned to the polysilicon gate instead of butting against the LOCOS edges. This test structure simulates the real CMOS inverter layout shown in Fig. 7.22,[17] although the actual p^+ and n^+ are not along the same direction of the polysilicon gate. When the gate is biased at 5 V, the n-channel surface under the polysilicon gate (indicated by n-CH) is inverted to n-type but is isolated from the n-well by the p-substrate as long as the field region in the p-substrate is not inverted. Similarly, as the gate is turned to 0 V, the p-channel surface under the polysilicon gate (indicated by p-CH) is inverted to p-type but is isolated to the p-substrate unless the field region inside the n-well is inverted.

The present test structure serves as a good vehicle to test the field inversion problem under the polysilicon lines. However, the true source and drain diffusions in the polysilicon gate case are kept further apart than in the aluminum gate case (Fig. 7.21). This instance tends to inhibit punchthrough and hence does not test the worst punchthrough problem.

As n^+-to-n-well or p^+-to-p-substrate spacing is reduced, the magnitude of the field threshold voltage for either nMOS or pMOS parasitic FET decreases, again similarly to the short-channel effect commonly observed in an active MOSFET. This threshold roll-off effect in parasitic asymmetric FETs is very serious because of the associated deep drain (n-well or p-substrate). The deep drain is also more likely to cause drain-to-source punchthrough, which is another failure mode in CMOS isolation. Moreover, the two failure modes, field inversion and punchthrough, are coupled through 2-D effects.[17] For worst-case considerations, field inversion must be measured at maximum drain bias and punchthrough must be measured at highest gate voltage.

Modelling. Two-dimensional process and device modelling is carried out using the SUPRA and GEMINI programs discussed in Chapter 3. The input parameters for the SUPRA program are adjusted to achieve agreement with the measured profiles in the vertical direction. The GEMINI program

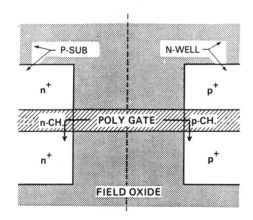

Figure 7.22 Layout top view in CMOS isolation region between n- and p-channel active devices (Ref. 17, © 1986 IEEE).

is basically a Poisson's equation solver, but is quite adequate for isolation modelling in which only very low levels of current are concerned.

Fig. 7.23(a) and (b) show the 2-D impurity distribution and the corresponding potential contours for the n^+-to-n-well aluminum gate asymmetric parasitic FET.[18]   The device is formed using a conventional diffused n-well technology.   The nominal separation of n^+ diffusion from the n-well (that

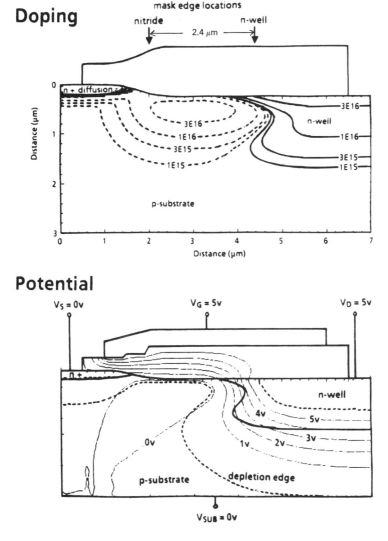

Figure 7.23   2-D simulated impurity doping and potential contours in the n^+-to-n-well parasitic MOSFET (Ref. 18, © 1987 IEEE)

is, the separation of the nitride LOCOS mask and the n-well implant resist mask) is 2.4 μm; the locations of the mask edges are indicated on Fig. 7.23(a).[18] Notice the fairly high p-type field doping level between the n^+ diffusion and the well.

Fig. 7.23(b) illustrates a solution to Poisson's equation for the structure depicted in Fig. 7.23(a). The n-well and parasitic gate are biased to 5 V with respect to the p-substrate and n^+ source; this instance represents the worst case for leakage with a 5 V supply. In this case, the parasitic FET is off as indicated by the 0 V surface potential in the middle of the channel. The suppression of lateral spread of the well-substrate depletion region by the p-type field dopants can be clearly seen.

Figs. 7.24(a) and (b) show simulated doping and potential contours, respectively for a p^+-to-p-substrate aluminum gate parasitic FET. In this case, a device formed using the retrograde n-well technology is taken as an example. The nominal separation of p^+ diffusion from the p-substrate (that is, the separation of the nitride LOCOS mask and the n-well implant resist mask) is 1.2 μm; the locations of the mask edges are indicated on Fig. 7.24(a).[18] Note that the retrograde well is shallower under the field oxide than under the active area,[19,20] and also that the elimination of an n-well drive-in again limits lateral diffusion of the well dopants. The solution to Poisson's equation illustrated in Fig. 7.24(b) is obtained assuming that the p-substrate and parasitic gate are biased to -5 V with respect to the n-well and p^+ source, again representing the worst case for leakage. The potential contours at 0 V or below are all well inside the oxide, indicating that the Si surface underneath is at positive voltage. This indication means that sufficient voltage margin exists before inversion occurs. Also, notice that the depletion region associated with the drain junction is mostly extended to the p-substrate rather than the n-well, which is much more heavily doped than the p-substrate. Because the n-well serves as the channel region in this structure and only a small portion is taken by the drain depletion region, lateral punchthrough between p^+-to-p-substrate is less likely to happen.

n^+-to-n-well parasitic FETs. Fig. 7.25(a) shows field inversion voltages defined by drain currents of 10 pA, 1 nA, and 0.1 μA, measured at a 5 V drain voltage for both aluminum and polysilicon gate n^+-to-n-well parasitic FETs.[18] The threshold lowering effect below 2 μm nominal n^+ to n-well separation is clear. The n-well in these devices acts like a deep drain, and the associated depletion region extends into the substrate toward the n^+ source. This extension causes short-channel threshold lowering and drain-induced barrier lowering similar to normal active devices.

The thresholds of the thinner gate oxide of the polysilicon gate parasitic MOS transistors are lower than those of the aluminum gate devices, and the curves corresponding to the different subthreshold drain currents are closer together due to the improved subthreshold swing. As described in Chapter

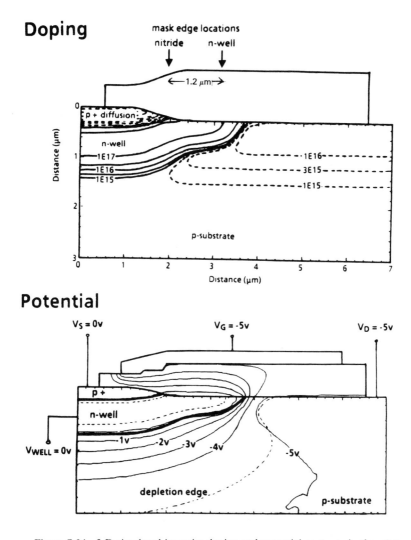

Figure 7.24   2-D simulated impurity doping and potential contours in the p^+-to-p-substrate parasitic FET (Ref. 18, © 1987 IEEE).

2, the subthreshold swing (S_t) decreases as the gate oxide is thinned.   More-over, threshold lowering is less for polysilicon gate FETs, which is again due to the thinner gate oxide.   At the smallest n^+-to-n-well separations (1–1.5 μm), the larger threshold reduction in the aluminum gate FETs causes their thresholds to be lower than those of polysilicon gate devices.   Consequently, it is the aluminum gate structures which ultimately limit the minimum sep-aration.   Furthermore, the oxide quality and thickness control of the alu-

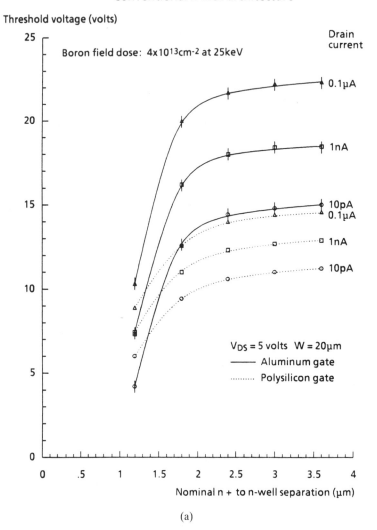

Conventional n-well architecture

Figure 7.25   (a) Threshold voltages; (b) breakdown voltages for the asymmetric n^+-to-n-well parasitic FETs (Ref. 18, © 1987 IEEE).

minum gate oxide often make its threshold less reproducible from run to run, and so more safety margin must be given.

The breakdown voltages of n^+-to-n-well parasitic FETs also indicate that the aluminum gate FETs ultimately determine minimum separation.   Fig. 7.25(b) shows that at 1.2 μm n^+ to n-well spacing, the polysilicon gate

Conventional n-well architecture

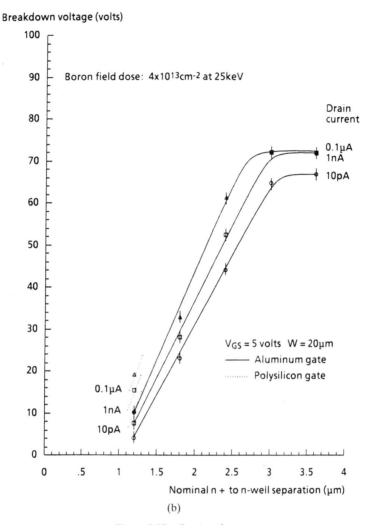

(b)

Figure 7.25 *Continued.*

punchthrough voltages measured at 10 pA, 1 nA, and 0.1 μA, with V_{GS} = 5 V are all about twice the corresponding values for aluminum gate devices. At larger separations, breakdown in the polysilicon gate devices is limited by the dielectric strength of the thin gate oxide, where the polysilicon overlaps the active area in the *n*-well (indicated as *p*-CH in Fig. 7.21(b)), and so punchthrough or avalanche breakdown cannot be measured. As compared with the symmetric parasitic FETs previously described, punchthrough in asymmetric FETs is much more severe because of the deeper drains in these

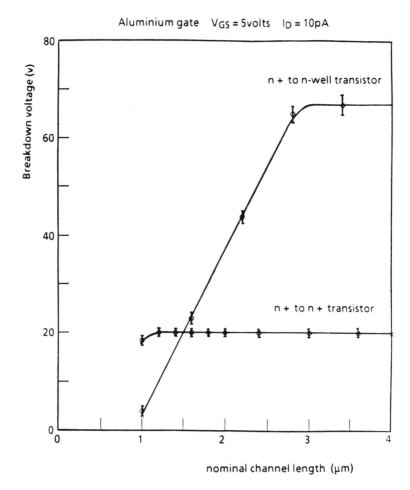

N-CHANNEL PARASITIC MOS TRANSISTOR BREAKDOWN VOLTAGES

Aluminium gate $V_{GS} = 5\,volts$ $I_D = 10pA$

Figure 7.26   Avalanche and punchthrough breakdown voltages for the symmetric (n^+-to-n^+) and asymmetric (n^+-to-n-well) parasitic nMOSFETs (Ref. 21, © 1987 IEEE).

devices.   Fig. 7.26 shows an obvious difference by comparing an n^+-to-n^+ parasitic FET to an n^+-to-n-well device.[21]

p^+-to-p-substrate parasitic FETs.   Field inversion voltages for p^+-to-p-substrate parasitic FETs are shown in Fig. 7.27 for both aluminum and polysilicon gate devices.[18]   Because nominal p^+-to-p-substrate separation represents the separation of the nitride LOCOS mask and the *n*-well resist mask, and the actual p^+ in the polysilicon device is abutted to the polysilicon edge rather than the nitride edge, the actual p^+ to *p*-substrate separation is larger than its nominal value by the length of the poly-overlapping thin oxide. Threshold values can only be obtained at 0.6 μm or less, again due to the dielectric breakdown in the thin oxide region. At nominal zero separation in the polysilicon devices, the actual p^+ to *p*-substrate separation is about the

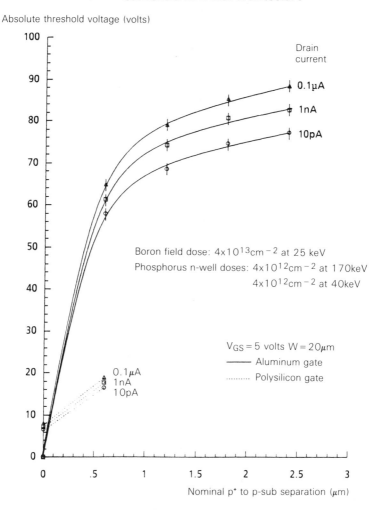

Figure 7.27 Threshold voltages for the asymmetric p^+-to-*p*-substrate parasitic FETs (Ref. 18, © 1987 IEEE).

same as the length of the poly-overlapping thin oxide, that is, 1.2 μm. For the aluminum gate case in which the p^+ is abutted to the LOCOS edge, the devices fail to operate.

p^+ to p-substrate lateral punchthrough is not observed until the nominal separation is reduced to 0.6 μm. As previously described, this minimum punchthrough is mostly due to the fact that the majority of the drain depletion region is extended into the lightly doped p-substrate, which acts as the drain in this structure. Vertical punchthrough between p^+ to p-substrate can be a more serious problem especially when the n-well depth is scaled down for CMOS VLSI. In general, vertical punchthrough presents a problem if leakage current flows between a p^+ and p-substrate through an n-well. In normal CMOS operation, the n-well is at V_{dd}, the p-substrate is at V_{ss} and the p^+ is biased between V_{dd} and V_{ss}. However, during transient overshoot, the p^+ or p-substrate can become negatively biased, so the condition should be considered.

The problem can be illustrated by using an example of a conventional but shallow (about 1 μm) n-well with its profile shown in Fig. 7.28(a).[22] The

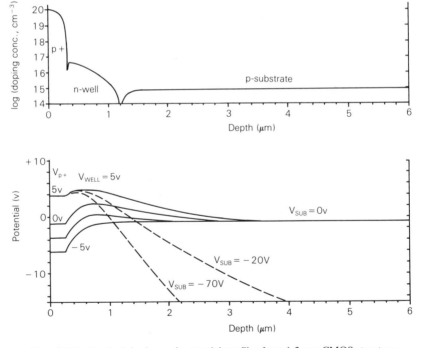

Figure 7.28 Vertical doping and potential profiles for a 1.2 μm CMOS structure. (a) Net doping, and (b) potential profiles, solid lines: p^+ voltage varies with p-substrate grounded, dashed lines: substrate bias changes with p^+ fixed at 5 V. The n-well is always at 5 V (Ref. 22, © 1987 IEEE).

corresponding potential distribution is shown in Fig. 7.28(b), in which the p^+ is biased between $+5$ V and -5 V while the n-well and p-substrate are biased at normal operation voltages shown as solid lines.[22]   As the p^+ is swept from 5 V to 0 V, the p^+/n-well depletion region extends further into the well; however, the potential barrier remains even when the well is fully depleted. Consequently, no vertical punchthrough current is present during normal operational conditions.   The barrier disappears when the p^+ potential reaches -5 V, thus punchthrough current flows from the substrate up to the p^+. Punchthrough current can also flow from the p^+ down to the substrate if it is negatively biased (shown as dashed lines).   Because most of the well-to-substrate depletion region is into the lightly-doped substrate rather than into the well, the p^+-to-substrate potential barrier cannot be lowered significantly until the substrate is at a large negative bias (< -70 V).   The situation is altered however if the p-substrate is replaced by a p-epi layer on a p^+ substrate.   Because the heavily-doped substrate prevents the extension of the well-substrate depletion region down to the substrate, the depletion region is forced up into the well.   As a result, the V_{SUB} needed to lower the potential barrier is reduced, thus making it easier to cause vertical punchthrough.   As shown in Fig. 7.29,[22] the V_{SUB} magnitude at the breakdown of vertical isolation

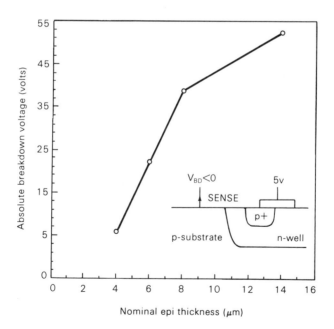

Figure 7.29 Vertical punchthrough voltage versus nominal (i.e., as-grown) epi thickness in a CMOS structure with 1.1 μm well depth. Sense current = 1 nA (Ref. 22, © 1987 IEEE).

decreases as the epi layer is thinned. At 4 μm nominal (as-grown) epi thickness, isolation breaks down at V_{SUB} of about -5 V. Therefore, vertical isolation may limit the use of the thinnest epi layer.

7.3.3 Retrograde Well Technologies to Improve Isolation

As discussed in Chapter 5, retrograde wells are formed by high-energy implantation after the LOCOS oxidation cycle. Lateral diffusion is greatly reduced resulting in a significant saving of layout space. Because a well(s) is basically used to separate n- and p-channel transistors, isolation among these opposite-type devices are expected to improve. The retrograde p-well technology[23] was originally proposed to reduce the well depth and corresponding layout area. Although this technology has shown sufficiently high field threshold voltages and p^+-to-p-well punchthrough voltages, n-well is used as an example to compare the differences between a retrograde and conventional well technology in CMOS isolation.[16,18] Moreover, the comparison will be restricted to aluminum gate structures.

To make a fair comparison, conventional n-wells in this comparison are formed by minimum diffusion to achieve well depths in the active area as shallow as the retrograde wells made by high-energy implantation. The self-aligned feature of the n-well and field implants is preserved in a conventional well to achieve rigorous isolation at minimum n^+ to p^+ spacings. In other words, the conventional well technology is optimized to compete with the retrograde-well approach.

The important process steps in forming the n-wells for both technologies are shown in Fig. 7.30.[18,20] Both technologies start with the definition of active areas by nitride masking. Next, the n-well is aligned to it. Blanket boron field implant is used to form a p-type channel stop for n-channel transistors. The boron impurity concentration inside the n-well is compensated by the phosphorus implant used for the n-well formation. The major difference is that a conventional well is formed prior to LOCOS whereas the formation of the retrograde well is done after LOCOS. Because a retrograde well is formed by high-energy ions implanting through the field oxide, the well depth in the field region is only about half of the well depth in the active region. This depth is ideal in the sense of providing a shallow drain for the n^+-to-n-well parasitic FET, thereby improving isolation. Meanwhile the well in the active area is kept relatively deep so that the performance of the active transistors in the well are not significantly degraded. Furthermore, in the active area, the implanted impurity profile peaks near the bottom of the well, which is desirable for vertical isolation. At the same time, the profile peaks near the Si surface in the field region, which again is ideal for forming a channel stop and for preventing p^+-to-p-substrate lateral punchthrough.

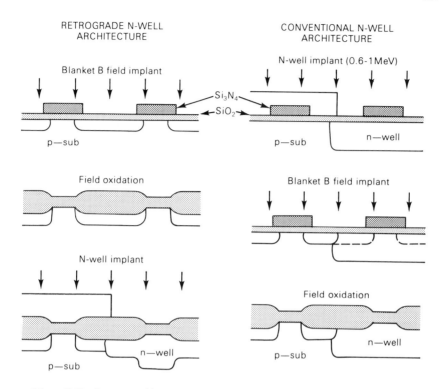

Figure 7.30   Process architectures for retrograde and conventional *n*-well CMOS structures (Ref. 16).

Figs. 7.31(a) and (b) show measured and modelled aluminum gate field inversion voltages and breakdown voltages, all defined at 10 pA drain current, for n^+ to *n*-well parasitic transistors with two different *n*-well technologies.[18] At large n^+ to *n*-well separations, avalanche breakdown at the well-substrate junction occurs before punchthrough takes place. The lower avalanche breakdown voltage observed with retrograde wells is attributed to higher well doping. As previously described, shallower retrograde wells should have less short-channel effects in terms of threshold lowering and drain-induced barrier lowering. These expectations are confirmed from the threshold results (Fig. 7.31(a)) and punchthrough data (Fig. 7.31(b)).

Fig. 7.32(a) shows field inversion voltages for lateral p^+-to-*p*-substrate parasitic devices.[18] The absolute values are much higher than those for n^+-to-*n*-well devices. These higher values are due in part to the higher field doping needed in the *n*-well to ensure adequate compensation of the boron. They are also partly due to the fixed charge at the Si–SiO$_2$ interface, which is likely to be positive and therefore increase the magnitude of the threshold voltage of *p*-channel transistors. It is also clear that short-channel effects

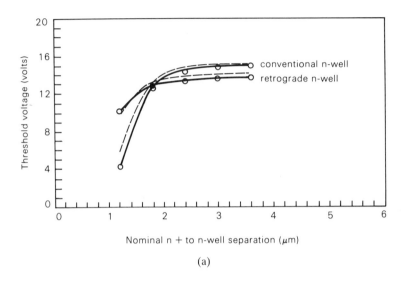

(a)

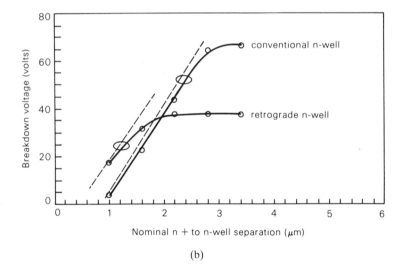

(b)

Figure 7.31   Comparison of retrograde and conventional well structures for n^+ to n-well parasitic FETs: (a) field threshold voltage; and (b) n^+ to n-well breakdown voltage. Solid lines: experimental data, dashed lines: modelled results, breakdown voltage is modelled by punchthrough only (Ref. 18, © 1987 IEEE).

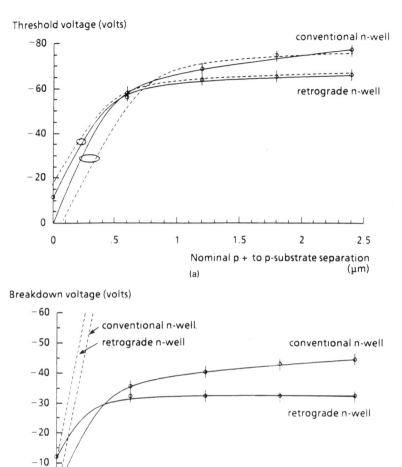

Figure 7.32 Comparison of retrograde and conventional well structures for p^+ to p-substrate parasitic FETs: (a) field threshold voltage; and (b) p^+ to p-substrate breakdown voltage. Solid lines: experimental data, dashed lines: modelled results, breakdown voltage is modelled by punchthrough only (Ref. 18, © 1987 IEEE).

are much less severe in these p-channel parasitic devices, particularly for the retrograde structures; the thresholds do not fall until the channel length reaches about 0.5 μm.   This effect is due to the higher channel (n-well) doping in these parasitic transistors, and to the drain (substrate) doping being lower than the channel (n-well) doping.

Fig. 7.32(b) shows breakdown voltages obtained with the same p^+ to p-substrate devices, and comparison with Figs. 7.31(a) and (b) again shows the superior performance of these structures.[18]   In these types of parasitic FETs, the p-substrate, which is regarded as the drain, is not only deep but wraps around under the p^+ source (Fig. 7.21).   Thus, the possibility of both lateral and vertical leakage paths between the p^+ diffusion and the p-substrate exists.   The limiting values of breakdown voltage for retrograde n-well structures is set by well to substrate avalanche breakdown, as in the case of the n^+ to n-well transistors.   However, with conventional wells, the limit is set by p^+-to-p-substrate vertical punchthrough breakdown.   Thus, the maximum breakdown voltage shown on Fig. 7.32(b) for these devices is not the same

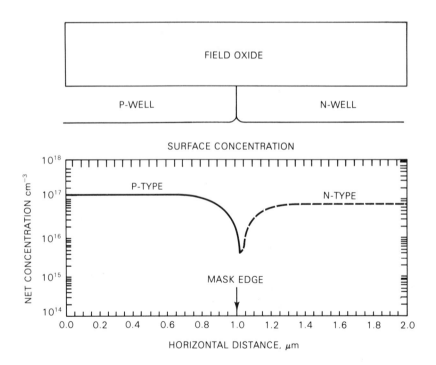

Figure 7.33   Surface impurity concentration at the boundary of the p- and n-retrograde wells in quadruple-well CMOS (Ref. 24, © 1984 IEEE).

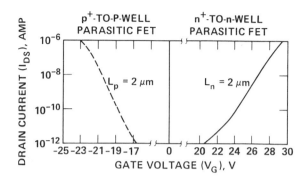

Figure 7.34   Subthreshold characteristics of asymmetric parasitic FETs in quadruple-well CMOS (Ref. 24, © 1984 IEEE).

as that shown on Fig. 7.31(b), even though the drain-channel junction is in fact the same in both cases.

The use of retrograde wells for both n- and p-wells has been applied to twin-tub CMOS technology and was called quadruple wells to emphasize the fact that for each type (n- or p-well), two wells were needed, one in the active area and another, shallower one in the field region. The two shallow wells serve as self-aligned channel stops with the n- and p-type impurity concentration tightly restricted on different sides of the well boundary. Fig. 7.33 shows that the reduction of net impurity concentration occurs only within a few tenths of a micron near the mask edge rather than a few microns, as observed in a conventional twin-tub structure (Fig. 7.18). Experimental data show that at a 4 μm n^+-to-p^+ separation, subthreshold leakage in the isolation region is <1 pA for gate voltages between -15 V and 20 V (Fig. 7.34).[24] Modelling results suggest that a 2 μm n^+-to-p^+ separation is sufficient for providing adequate isolation.[17]

In spite of all the advantages associated with the retrograde well in device isolation, specific concerns for this technology have been expressed. As discussed in Chapters 5 and 6, the active devices made inside a retrograde well generally suffer more from junction capacitance and body effect. The former leads to a decrease in speed of about 5–10% in a typical CMOS logic circuit. Another major concern is the availability of the high-energy implanter and the lack of understanding in using such a machine in a production environment.

7.4 NEW ISOLATION TECHNIQUES FOR CMOS

7.4.1 Trench Isolation

Trenches in the Si substrate have been proposed for CMOS isolation.[25,26] This isolation technique has several advantages. First, it eliminates oxidation encroachment (known as bird's beak) in LOCOS. The associated narrow-channel effect is then prevented. Second, trenches filled with oxide provide a planar surface, thus avoiding a step-coverage problem. Finally, when trenches of several microns depth are used in a CMOS chip, they provide a physical separation between opposite-type devices. Consequently, isolation can be achieved with very narrow n^+-to-p^+ spacing. Fig. 7.35(a) shows the schematic of an n-well CMOS structure with trenches which can be as narrow as 1 μm separating the n^+ and p^+. Notice that the trenches here are deeper than the n-well to effectively isolate the p-channel device.

Trench isolation also has several problems associated with it, making this technology difficult in production. First, fabrication is very complicated

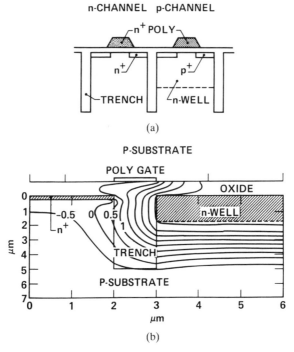

(a)

(b)

Figure 7.35 Trench isolation and potential distribution. (a) Cross-sectional view of an n-well CMOS with trench isolation; (b) the corresponding 2-D equipotential contours; the simulation is done at $Q_{fc} = 5 \times 10^{10}$ cm^{-2}, $N_A = 6 \times 10^{14}$ cm^{-3}, $V_{well} = 3$ V, $V_G = 0$ V, $V_{sub} = -1$ V (Ref. 27, © 1983 IEEE).

because it involves anisotropic etching, conformal oxide or poly deposition, and planarization. All these processes become even harder if trenches with variable widths are needed. Second, process-induced damage and deposited films always produce fixed charges (Q_{fc}) inside and at the boundary of the trenches. Hence, extra leakage appears along trench sidewalls. Finally, sidewall inversion is a major problem that can place a limitation in layout. Sidewall inversion is caused by the horizontal parasitic MOS device with the well acting as a gate electrode and the trench dielectric as MOS gate oxide with thickness equal to the trench width. The voltage across this MOS device is 5 V under normal CMOS operating conditions. This 5 V gate voltage, and the narrow (e.g., 1 μm) trench width can easily cause inversion at the sidewall outside, but facing the well. In the case of n-well technology, the situation is worse because both lightly-doped p-substrate and Q_{fc} (normally positive) escalate sidewall inversion. The 2-D potential contours in Fig. 7.35(b) indicate that the surface potential at the sidewall opposite the n-well exceeds 0.5 V, causing inversion there.[27] Once sidewall inversion occurs, n-channel devices with n^+ butted to the same sidewall will short along the sidewall. An obvious solution is to leave a separation between the n^+ and the sidewall or to increase trench width; however, both are done at the expense of layout space, thereby defeating the purpose of this technology. More appropriate approaches would be to reduce Q_{fc} through better control in trench etching and filling, and/or to dope the sidewall surface to a higher concentration, i.e., forming the channel stop at the sidewall. The former approach is difficult because it requires extensive process refinements and detail characterization of Q_{fc} at the sidewall. The latter approach is effective in preventing sidewall inversion as shown in Fig. 7.36.

In this example, the Si potential at the sidewall decreases from 0.5 V (Fig. 7.36a) to negative voltages (Fig. 7.36b) when the sidewall is boron-doped to 5×10^{17} cm^{-3}. Low subthreshold leakage has been measured from an n-channel MOSFET, with its source and drain abutted to the sidewall.[28] However, it is extremely difficult to implement a CMOS structure in which only one side of the trench is doped. In the n-well example shown in Fig. 7.35a, only the sidewall in the p-substrate needs to be doped to form a p-type channel stop, the other side of the trench is in the n-well and should not be counter-doped by p-type impurity atoms. The payoff for trench isolation technology in CMOS may not be worth the extra complexity and associated cost[29] unless minimum n^+-to-p^+ separation is absolutely critical to achieve very high packing density.

7.4.2 Selective Epi Growth

Selective epitaxial growth (SEG) has recently been developed for CMOS isolation.[30,31] In this technology, Si is first etched away from the well areas using an oxide mask, then a layer of insulator is formed at the sidewall, and

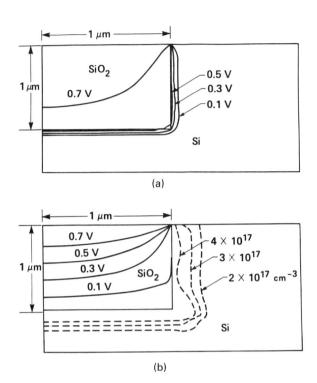

Figure 7.36 2-D potential contours simulated for a trench structure in a p-type substrate with $N_A = 8 \times 10^{16}$ cm^{-3}: (a) no sidewall doping; (b) with sidewall doping. Dashed lines correspond to the doping contours at the sidewall.

finally the well is filled by selective epi growth. Fig. 7.37 shows the schematic of the processing steps for an n-well CMOS. Here phosphorus ions (P^+) are implanted after the sidewall oxide is formed.[31] The SEG technique is then used to grow Si only on the Si substrate, but not on the oxide surface. The n^+ at the bottom improve vertical isolation and lower well sheet resistance. With this etch-and-regrowth technique, doping sidewalls outside the n-well are implemented by depositing a boron-doped film at the sidewall and driving boron into the sidewall before the formation of the sidewall insulator. It is also advantageous to dope the sidewall inside the n-well to higher n-type concentration, but it is not necessary because the n-well is already highly doped and, with the positive Q_{fc}, the n-type Si next to the trench oxide (see arrows in Fig. 7.37d) is less likely to invert. In this process, LOCOS is still needed to form field oxide with various isolation widths.

One way to eliminate LOCOS completely is illustrated in Fig. 7.38.[31] A thick oxide layer is formed and windows are etched in the oxide using the active-area mask. Selective epi growth is then applied to fill the windows followed by an n-well diffusion. The sidewall doping is difficult to achieve

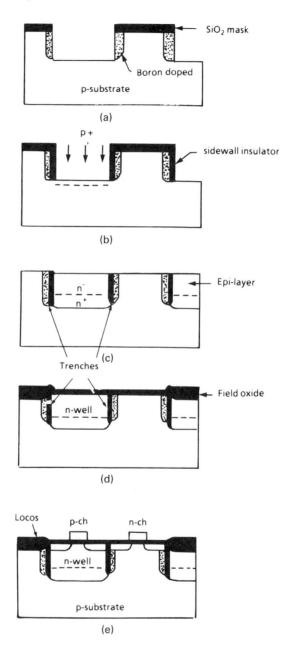

Figure 7.37   SEG process steps for making CMOS with trench and LOCOS isolation (Ref. 31).

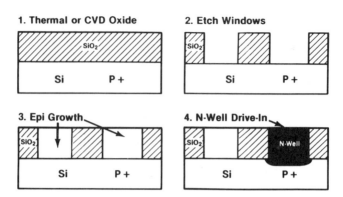

Figure 7.38 SEG process steps for making CMOS with variable isolation widths (Ref. 31).

with this method. A technique to achieve both sidewall doping and variable oxide widths can be done by applying SEG twice to form *n*- and *p*-wells independently. As shown in Fig. 7.39,[31] sidewall doping can be added immediately after Step 5 by depositing a boron-doped film, filling *p*-type SEG epi, and driving boron atoms into *p*-well sidewalls.

Although the SEG technology appears attractive in CMOS isolation, it is still in its infancy. More work needs to be done to understand material quality at the sidewall of the SEG growth epi layer. Obviously, good selectivity and a clean substrate are key issues in this technology.

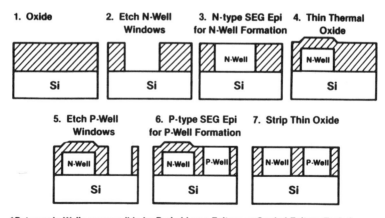

*Retrograde Wells are possible by Buried Layer Epitaxy or Graded Epitaxy Techniques

SiO₂

Figure 7.39 SEG process steps for independent *p*- and *n*-well formation (Ref. 31).

7.4.3 SOS and SOI Isolation

Silicon-on-insulator (SOI) CMOS structures eliminate all previously-mentioned isolation problems by building n- and p-type islands on an insulating substrate for p- and n-channel transistors, respectively. The SOS and SOI process technologies were discussed in Chapter 5, Sec. 5.8 and Sec. 5.9. Similar or opposite type transistors are isolated by insulator and possibly by air. Field inversion and punchthrough do not exist in an SOI or SOS technology, but new parasitic transistors inherent in an SOS or SOI structure can produce leakage current.

As shown in Fig. 7.40(a), a silicon island is formed on an insulator substrate such as sapphire or oxide.[32]   A thin gate oxide is grown and the polysilicon line crossing-over the island serves as the gate of the MOS transistor. Two parasitic devices can be found in this structure. First, the substrate contact, the insulating substrate, and the backside of the island form a Metal-Insulator-Semiconductor (MIS) FET, referred to as a back-channel MISFET. Second, the polysilicon gate over the island sidewall forms a sidewall parasitic FET in parallel with the active MOS transistor. Both parasitic devices can result in leakage current, although the leakage is intra-device rather than inter-device in its nature. Because defect density at the Si island and insulating substrate is normally high due to lattice mismatch and film contamination, a back channel is easily formed because of inversion. Current flowing in this back channel cannot be switched off by applying voltage on the poly gate at the front surface. Therefore, a residual off-state leakage current is normally observed as shown in Fig. 7.40(b).[32]   Another leakage current, edge leakage, is also shown in the figure. This leakage component arises from the parasitic sidewall FET. For a $\langle 100 \rangle$ SOS or SOI film, crystal orientation at the sidewall is $\langle 111 \rangle$ which normally results in higher fixed oxide charge (Q_{fc}) after oxide growth. Consequently, threshold voltage at the sidewall is lower than at the active transistor, thereby resulting in a leakage component in the subthreshold characteristics of the active device. Higher leakage reduces noise margin and increases power dissipation in CMOS circuits built on SOS or SOI technology.

7.5 ISOLATION DESIGN RULES IN CMOS

The results shown in Fig. 7.31 indicate that it is possible to maintain adequate isolation between n^+ diffusions and n-wells at about 1.5 μm nominal separation by using conventional wells, and down to 1 μm separation with retrograde wells. At the same time, the results in Fig. 7.32 show that p^+-to-p-substrate separations down to 0.6 μm can be achieved with either well type. Thus it should be possible to realize full CMOS structures with about 2 μm n^+-to-p^+ separation using conventional n-wells, and about 1.6 μm separation with a retrograde well technology. However, these results are obtained with

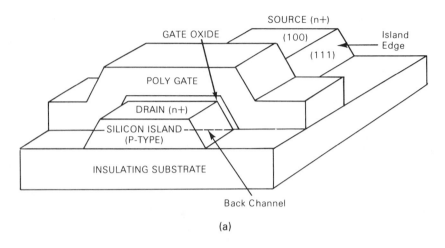

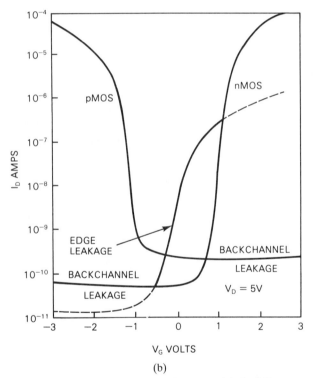

Figure 7.40 SOS and typical electrical performance: (a) the 3-D representation of an SOS island structure; (b) the corresponding subthreshold characteristics showing edge and back channel leakage for both *n*MOS and *p*MOS (Ref. 32).

isolated n^+-to-n-well and p^+-to-p-substrate parasitic transistors. To establish isolation design rules, it is necessary to demonstrate such small separations in full CMOS devices.

Fig. 7.41 shows supply currents drawn by full CMOS structures with 1.8 μm n^+-to-p^+ separations (1.2 μm n^+-to-n-well edge and 0.6 μm p^+-to-n-well edge).[18] Fig. 7.41(a) shows the measurement arrangement. The output node, connecting the n^+ and p^+ diffusions closest to the well edge, is either shorted to ground ($V_{OUT} = 0$) or to the supply line ($V_{OUT} = V_{DD}$). This instance simulates an inverter with its output in either a Low or High state. Ideally the supply current drawn by the CMOS device should be zero in both cases, but any leakage due to inadequate isolation between the n- and p-channel devices will be detected as a component of the supply current.

Fig. 7.41(b) shows results for conventional well devices. When $V_{OUT} = V_{DD}$, no potential difference exists between the n^+ diffusion and the n-well, so no leakage current flows between them. The limiting mechanism in this case is p^+ to p-substrate breakdown, and according to Fig. 7.32, no breakdown should be observed below 20 volts. Clearly, the worst case for leakage is with the output node at ground. Under this condition, the full supply voltage appears between the n^+ diffusion and the n-well, and the mechanism limiting the maximum operating voltage is n^+-to-n-well punchthrough leakage. The results shown in Fig. 7.41(b) for $V_{OUT} = 0$ are consistent with the punch-through breakdown voltages shown in Fig. 7.31(b). Fig 7.41(b) shows that significant leakage occurs at $V_{DD} = 5$ volts or above. The CMOS device shown does not provide isolation with sufficient safety margin for 5 V operation. Increasing the n^+ to n-well spacing to 1.8 μm (n^+ to p^+ separation

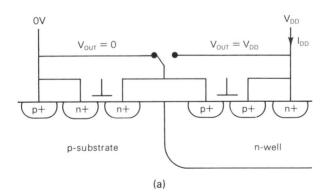

(a)

Figure 7.41   Test structures and I-V curves for a full CMOS isolation structure: (a) test structure and measurement arrangement; (b) leakage current for conventional well; and (c) leakage current for retrograde well devices (Ref. 18, © 1987 IEEE).

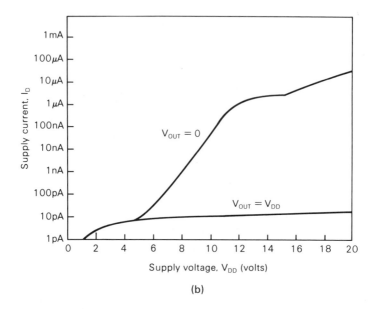

(b)

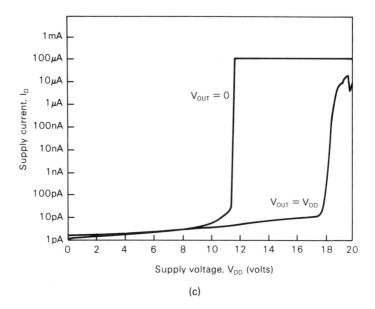

(c)

Figure 7.41 *Continued.*

= 2.4 μm) gives a maximum operating voltage limited by junction breakdown (just over 20 volts, in this case).

Fig. 7.41(c) shows results for retrograde well devices with 1.8 μm n^+-to-p^+ separation. In these cases, the leakage current remains below 10 pA for supply voltages as high as 10 V, providing a good safety margin for 5 V operation. Again, the condition with $V_{OUT} = 0$ sets the lowest breakdown voltage, although p^+-to-n-well junction avalanche breakdown occurs at only a few volts more than the onset of n^+-to-n-well punchthrough leakage.

In a standard LOCOS-isolated CMOS IC, isolation among similar-type transistors can be achieved with 1 μm isolation spacing. Isolation among opposite-type transistors is more involved, however modelling and measurement results show that it is possible to achieve isolation between an n- and a p-channel transistor with approximately 2.4 μm isolation spacing. The use of retrograde wells can further reduce this isolation dimension to below 2 μm. At such small isolation dimensions, latchup becomes a major problem and ultimately may be the factor limiting the isolation spacing between an n- and a p-channel transistor. Latchup in CMOS will be discussed in the next chapter.

REFERENCES

1. K. L. Wang, S. A. Sadler, W. R. Hunter, P. K. Chatterjee, and P. Yang, "Direct moat isolation for VLSI," *IEEE Trans. Electron Devices*, vol. ED-29, pp. 541–547, 1982.

2. S. M. Sze, Physics of Semiconductor Devices, 2nd ed. New York: Wiley, 1981.

3. T. V. Nordstrom and C. F. Gibbon, "The effect of gate oxide thickness on the radiation hardness of silicon-gate CMOS," *IEEE Trans. Nucl. Sci.*, vol. NS-28, pp. 4349–4353, 1981.

4. E. Kooi, J. G. Van Lierop, and J. A. Appels, "Formation of silicon nitride at a Si–SiO₂ interface during local oxidation of silicon and during heat treatment of oxidized silicon in NH₃ gas," *J. Electrochem. Soc.*, vol. 123, p. 1117, 1976.

5. R. A. Monline and G. W. Reutlinger, "Self-aligned maskless chan-stops for IGFET integrated circuits," *IEEE Trans. Electron Devices*, vol. ED-20, pp. 1129–1132, 1973.

6. J. Y. Chen and R. C. Henderson, "Radiation effects of *e*-beam fabricated submicrometer NMOS transistors," *IEEE Electron Device Lett.*, vol. EDL-3, pp. 13–15, 1982.

7. W. G. Oldham, "Isolation technology for scaled MOS VLSI," in *IEDM Tech. Dig.*, pp. 216–219, 1982.

8. J. Y. Chen, R. C. Henderson and D. E. Snyder, "A novel self-aligned isolation process for VLSI," *IEEE Trans. Electron Devices*, vol. ED-30, p. 1521, 1983.

9. J. Hui, T. Y. Chiu, S.-W.S. Wong, and W. G. Oldham "Sealed-interface local oxidation technology," *IEEE Trans. Electron Devices*, vol. Ed-29, pp. 554–561,

1982; also, J. Hui, T. Y. Chiu, S. Wong, and W. G. Oldham, "Selective oxidation technologies for high density MOS," in *IEEE Electron Device Lett.*, vol. EDL-2, p. 244, 1981.

10. K. Y. Chiu, J. L. Moll, and J. Manoliu, "A bird's beak free local oxidation technology feasible for VLSI circuits fabrication," *IEEE Trans. Electron Devices*, vol. ED-29, pp. 536–540, 1982; K. Y. Chiu, R. Fang, J. Lin, and J. L. Moll, "The SWAMI—A defect free and near-zero bird's beak local oxidation technology for VLSI," in *Proc. 1982 Symp. VLSI Technol.*, pp. 28–29, 1982.

11. K. Kurosawa, T. Shibata, and H. Izuka, "A new bird's beak free field isolation technique for VLSI Devices," in *IEDM Tech. Dig.*, pp. 384–387, 1981.

12. K. Ehara, T. Morimoto, S. Muramoto, T. Hosoya, and S. Matsuo, "Perfect planar technology for VLSIs," in *Proc. 1982 Symp. VLSI Technol.*, pp. 30–31, 1982.

13. J. Y. Chen, R. C. Henderson, J. T. Hall, and E. M. Yee, "A fully recessed field isolation technology using photo-CVD oxide," in *IEDM Tech. Dig.*, pp. 233–236, 1982.

14. L. C. Parrillo, R. S. Payne, R. E. Davis, G. W. Reutlinger, and R. L. Field, "Twin-tub CMOS: A technology for VLSI circuits," in *IEDM Tech. Dig.*, p. 752, 1980; also L. C. Parrillo, L. K. Wang, R. D. Swenumson, R. L. Field, R. C. Melin, and R. A. Levy, "Twin-tub CMOS II: An advanced VLSI technology," in *IEDM Tech. Dig.*, p. 706, 1982; also, J. Agraz-Guerena, R. Ashton, W. Bertram, R. Melin, R. Sun and J. T. Clemens, "Twin Tub III—A third generation CMOS technology," in *IEDM Tech. Dig.*, p. 63, 1984.

15. J. Y. Chen, "CMOS—The Emerging VLSI Technology," *IEEE Circuits and Devices*, vol. 2, p. 16, March 1986.

16. A. G. Lewis, R. A. Martin, and J. Y. Chen, "Retrograde and conventional *n*-well CMOS technologies: A comparison," Internal Report, EIL-87-17, Xerox Palo Alto Research Center, Oct. 1986.

17. J. Y. Chen and D. E. Snyder, "Modeling Device Isolation in High Density CMOS," *IEEE Electron Device Lett.*, vol. EDL-7, p. 64, 1986.

18. A. G. Lewis, J. Y. Chen, R. A. Martin, and T. Y. Huang, "Device Isolation in High Density LOCOS-Isolated CMOS," *IEEE Trans. Electron Devices*, vol. ED-34, p. 1337, June 1987.

19. R. A. Martin and J. Y. Chen, "Optimized Retrograde *N*-Well for 1-μm CMOS Technology," *IEEE J. of Solid State Circuits*, vol. SC-21, p. 286, 1986.

20. R. A. Martin, A. G. Lewis, T. Y. Huang, and J. Y. Chen, "A new process for one micron and finer CMOS," *IEDM Tech. Dig.*, p. 403, 1985.

21. J. Y. Chen and A. G. Lewis, "Parasitic transistor effects in CMOS VLSI," *IEEE Circuits and Devices Magazine*, Vol. 4, p. 8, May 1988.

22. A. G. Lewis, R. A. Martin, J. Y. Chen, T.-Y. Huang, and M. Koyanagi, "Vertical isolation in shallow *n*-well CMOS circuits," *IEEE Electron Device Lett.*, EDL-8, p. 107, 1987.

23. R. D. Rung, C. J. Dell'Oca, and L. G. Walker, "A retrograde *p*-well for high-density CMOS," *IEEE Trans. Electron Devices*, vol. ED-28, p. 1115, 1981.

24. J. Y. Chen, "Quadruple-Well CMOS for VLSI Technology," *IEEE Trans. Electron Devices*, vol. ED-31, p. 910, 1984.

25. R. D. Rung, H. Momose and Y. Nagakubo, "Deep trench isolated CMOS devices," *IEDM Tech. Dig.*, p. 6, 1982.

26. T. Yamaguchi, S. Morimoto, G. Kawamoto, H. K. Park, and G. C. Eiden, "High-speed latch-up-free 0.5 μm-channel CMOS using self-aligned Ti-Si and deep-trench isolation technologies," in *IEDM Tech. Dig.*, p. 522, 1983.

27. K. M. Cham, S. Chiang, D. Wenocur, and R. D. Rung, "Characterization and modelling of the trench surface inversion problem for the trench isolated CMOS technology," in *IEDM Tech. Dig.*, p. 23, 1983.

28. G. Chang, private communication.

29. R. D. Rung, "Trench isolation prospects for application in CMOS VLSI," in *IEDM Tech. Dig.*, p. 574, 1984.

30. T. I. Kamins, S. Y. Chiang, D. R. Bradbury, and D. B. Rao, "A CMOS Device isolation structure using the selective-etch-and-refill-with-epi (SEREPI) Process," in *IEDM Tech. Dig.*, p. 757, 1985.

31. J. Borland, "Submicron CMOS/Epi Technology," presented in the Seminar on Innovations in VLSI/USLI CMOS Technology, Santa Clara, Dec. 5, 1986.

32. D. Leong, "Silicon-on-sapphire," in the Lecture Notes of the short course on Silicon-on-Insulator (SOI) Technologies for Integrated-Circuit Applications, U.C. Berkeley, Oct. 1983.

EXERCISES

1. Suppose the field oxide under a n^+ polysilicon line is 6000 Å and the p-type channel stop under the oxide is doped to 1×10^{17} cm^{-3}, calculate the threshold voltage and the subthreshold swing factor if the oxide charge density is 1×10^{11} cm^{-2}.

2. If the measured threshold voltage of an aluminum-gate field oxide test device is 8 V and the measurement is done at 1 μA, what is the leakage current when the aluminum line is at 5 V (assuming the subthreshold swing is 0.5 V/dec at room temperature)? If the threshold voltage decreases to 6.5 V when temperature is increased to 125°C, what is the leakage current at 5 V?

3. Based on the I-V relation of an MOSFET, measured current is proportional to the electric channel width (W_e). Describe how one can extract W_e by measuring a series of test devices with different drawn channel widths.

4. Using a planar isolation technology such as fully-recessed LOCOS, assume that electrical channel narrowing effect is negligible. Supposing the difference between the actual physical channel width and the drawn channel width is 0.8 μm and the transconductance of a minimum size FET must be at least 0.7 mS, what should be the minimum drawn channel width? If the physical separation between two adjacent FETs must be 2 μm to provide proper isolation, what should be the minimum drawn separation? And what is the minimum drawn and actual pitch in the active area? (Assume that the FET has $t_{ox} = 200$ Å, $m_n = 500$ cm²/V-sec, $L_{eff} = 1$ μm.)

5. Based on Figs. 7.19 and 7.20, obtain the subthreshold swing (S_t) for the parasitic n- and p-channel FETs (both Al- and poly-gate devices) with a 2 μm separation between the two n^+ or two p^+ diffusion regions.

6. Your circuit designer wants to have no more than 10 pA leakage between the n^+ and n-well in his CMOS design and you have obtained the device and process data as shown in Fig. 7.25(a). What design rule should you provide for the minimum n^+ to n-well separation?

7. In a (p^+ diffusion)-(n-well)-(p-on-p^+ substrate) structure, the n-well is uniformly doped at 1×10^{16} cm^{-3} and is 1 μm deep, the p-epi thickness is 4 μm, it is uniformly doped at 1×10^{15} cm^{-3}, and the p^+ diffusion is 0.4 μm deep. If a substrate generator is needed and the p^+ diffusion and n-well are both biased at 5 V, what is the maximum (in absolute value) substrate voltage that can be applied before punchthrough occurs?

8. Comparing retrograde well to conventional well, what is the most important difference in process sequence and finished device structure? How does it affect device isolation?

9. In an SOS structure, how much substrate voltage must one apply at the back of the 250 μm-thick wafer to compensate an increase of 1×10^{10} cm^{-2} fixed charge density at the sapphire–Si interface so that back channel leakage can be avoid?

8

LATCHUP IN CMOS

This chapter will present another major problem in CMOS, latchup. This problem, like the isolation problem described in the last chapter, is also a strong function of circuit density, nMOS-to-pMOS separation in particular. Both device technologists and circuit designers should have knowledge of this subject to develop CMOS processes and lay out CMOS circuits.

8.1 INTRODUCTION

A bulk CMOS structure has inherent pnp and npn parasitic bipolar transistors formed by the substrate, the oppositely doped well and an active source/drain diffusion in either the well or the substrate. They are different from the active bipolar devices in a BiCMOS structure. The collector of either parasitic transistor is connected to the base of the other, forming a $pnpn$ parasitic SCR (Semiconductor Controlled Rectifier). Under certain conditions such as terminal overvoltage stress, transient displacement currents or ionizing radiation, lateral currents in the well and substrate can cause sufficient ohmic drop, hence forward biasing emitter-base junctions and activating both bipolar devices. When the current gain product of the two bipolars is sufficient to cause regeneration, the $pnpn$ SCR can be switched to a low impedance, high-current state. This condition is defined as latchup; and the condition is self-sustaining after the original stimulus is removed, provided the power supplies are capable of sourcing the excess current. Latchup, when it occurs in CMOS, can result in momentary or permanent loss of circuit function.

As CMOS feature sizes continue to shrink for VLSI circuits, lateral and vertical dimensions are scaled, resulting in better parasitic bipolar transistors. Consequently, latchup is a problem of increasing concern in VLSI, especially for circuits with n- and p-channel MOSFETs tightly packed together. The following sections will describe the condition for latchup to occur using a simple lumped equivalent circuit model. Then, individual components will be examined in the equivalent circuit. Characterization of latchup in both triggering and holding conditions will be discussed first in steady state, then in transient mode. Various methods of avoiding latchup, including layout techniques, will then follow.

8.2 PHYSICS AND LUMPED CIRCUIT MODEL

For the n-well CMOS shown in Fig. 8.1, the p^+, n-well and p-substrate form a parasitic vertical *pnp* bipolar device, whereas the n^+, p-substrate and n-well produce a parasitic lateral *npn* device. The simplest lumped-element equivalent circuit model is illustrated by the bold lines shown in the figure.

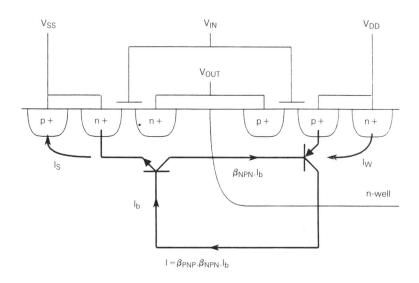

Figure 8.1 Schematic and equivalent circuit of parasitic bipolar transistors in a CMOS.

When the substrate current (I_S) and the well current (I_W) are sufficient to cause an ohmic drop of approximately 0.7 V, the base-emitter junctions for both transistors are sufficiently forward-biased ($V_{BE} \cong 0.7$ V) to provide significant current as described in Chapter 5, Sec. 5.10.2.   Bipolar devices are therefore in the active mode.   Because the base of the *pnp* connects to the collector of the *npn* and conversely, a feedback loop is formed by the two bipolar transistors with the loop gain equal to the betaproduct ($\beta_{pnp}\beta_{npn}$). β_{pnp} and β_{npn} are the common-emitter current gains for the *pnp* and *npn* transistors, respectively.   If the loop gain is less than one, I_b decays to zero as does I.   To have positive feedback, the loop gain must be greater than one.   This condition is necessary for latchup to occur.[1-3].

Consider the effect of well and substrate resistances.   The fact that the 0.7 V ohmic drop is produced is due to the finite resistances associated with the well and the substrate; they should be included in the circuit model, as shown in Fig. 8.2.   The *npn* collector current made of electrons ($I_b\beta_{npn}$) does not flow entirely into the *pnp* base because a part of it is shunted to the n^+ well contact through the well resistor (R_W).   This current is denoted as I_{RW}. The remaining current, $I_b\beta_{npn} - I_{RW}$, becomes the *pnp* base current.   Thus, the *pnp* collector current is $(I_b b_{npn} - I_{RW})\beta_{pnp}$ by definition.   Similarly, a portion of the *pnp* collector current (denoted as I_{RS}) is shunted to the substrate

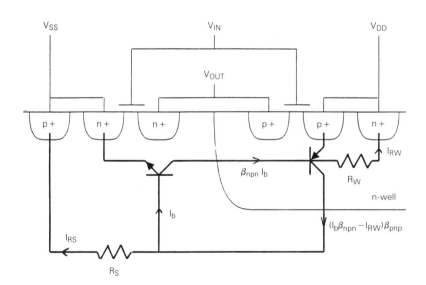

Figure 8.2   Schematic and equivalent circuit of parasitic resistors and bipolar transistors in a CMOS.

contact through the substrate resistance (R_S). The feedback current flowing into the *npn* base is now the *pnp* collector current offset by I_{RS}, i.e., $(I_b b_{npn} - I_{RW}) b_{pnp} - I_{RS}$. To cause positive feedback, this current must be greater than the initial *npn* base current I_b. This condition can be written as

$$(I_b \beta_{npn} - I_{RW}) \beta_{pnp} - I_{RS} > I_b \tag{8.1}$$

and rearranged as

$$I_b (\beta_{npn} \beta_{pnp} - 1) > I_{RS} + I_{RW} \beta_{pnp} \tag{8.2}$$

I_b can be expressed in terms of total supply current I_{dd} as

$$I_{dd} = I_{RS} + I_b (\beta_{npn} + 1) \tag{8.3}$$

which can be rearranged as

$$I_b = (I_{dd} - I_{RS})/(\beta_{npn} + 1) \tag{8.4}$$

Substituting Eq. 8.4 into Eq. 8.2, the following condition for latchup to occur can be obtained:

$$\beta_{npn} \beta_{pnp} > 1 + \frac{(\beta_{npn} + 1)(I_{RS} + I_{RW} \beta_{pnp})}{(I_{dd} - I_{RS})} \tag{8.5}$$

where I_{RS} and I_{RW} are approximately equal to V_{ben}/R_S and V_{bep}/R_W, respectively. V_{ben} and V_{bep} are the base-emitter voltages of the *npn* and the *pnp* transistors, respectively. They are approximately 0.7 V when the transistors are forward-biased into conduction.

The beta product $(\beta_{npn} \beta_{pnp})$ must now be greater than a value larger than one. This value increases as the resistor value $(R_S$ or $R_W)$ is decreased. Considering the worst case analysis, a beta product greater than one is indeed a necessary, but not sufficient, condition. To make the latchup requirement more difficult to meet, the value of the right-hand side of Eq. 8.5 can be increased through an increase of I_{RS} and/or I_{RW} by lowering R_W and/or R_S. Obviously, reducing β_{npn} and/or β_{pnp} always makes latchup more difficult by producing lower loop gain.

Although the lumped circuit model provides an understanding of the physics and condition for latchup to occur, circuit elements such as resistor values are difficult to estimate accurately. The following section will discuss characteristics of the transistors and values of the resistors.

8.3 PARASITIC TRANSISTORS AND RESISTORS

As previously discussed, there are two parasitic bipolar transistors: the vertical device and the lateral device. For an *n*-well CMOS structure, the vertical device is a *pnp* transistor and the lateral device is an *npn* device. When voltage drops caused by the well and substrate currents flowing along the parasitic series resistances R_W and R_S are about 0.7 V, the emitter-base junctions become sufficiently forward-biased, turning the two transistors on. Each transistor, once in the active mode, carries considerable current depending on the gain of the transistor. Moreover, the current can increase continuously due to the strong coupling between the two devices.

8.3.1 Vertical Bipolar Transistor

For a vertical bipolar transistor, current gain depends on well depth, well concentration and the built-in field in the well. The well depth sets the base width and the well concentration determines the Gummel number. For a typical 1 μm CMOS technology, the well is usually 1–2 μm deep, doped to 1×10^{16} cm^{-3}. For conventional n-well technology, the common-emitter current gain (β) of the corresponding *pnp* transistor with p^+ area of 8 μm × 40 μm is roughly 100. If a high-energy implant is used to form n-wells with a retrograde doping profile, a lower gain of approximately 10 is obtained due to the associated retarded built-in field and the higher doping concentration in the well. This β value is still too high to eliminate latchup susceptibility. If a p-well or a twin-tub with n-substrate technology is selected, the vertical *npn* transistor has a β value 2–3 times higher than the vertical *pnp* β due to higher minority carrier mobility in base region.

Transient response of the bipolar transistor is also important, because latchup in a real circuit is normally induced by transient triggering. Based on Eq. (5.15), the transit time for minority carriers across the base region can be extended to

$$\tau_{tr} = W_B^2/(\eta D) \tag{8.6}$$

where W_B is the base width, D is the minority carrier diffusion coefficient, and $\eta = 2$ for a uniformly doped base.[4] For a non-uniformly doped base, the associated internal built-in field results in an increase or decrease of the η value depending on whether it is an aiding or retarding field. The base transit time can be experimentally obtained by measuring the maximum frequency response[5] as described in Chapter 5. A typical τ_{tr} value for a 1-μm n-well CMOS technology is a few nanoseconds. The value can double if a retrograde well structure is used, due to the inherent retarding field.

8.3.2 Lateral Bipolar Transistor

For a lateral bipolar transistor, current gain strongly depends on the layout spacing between the diffusion outside the well to the well edge because the distance represents transistor base width. Field doping outside the well and the well depth are other parameters that also affect the lateral β. Because the base width here is generally larger than that of a vertical transistor, and the emitter and collector efficiencies are poor, the β value is normally an

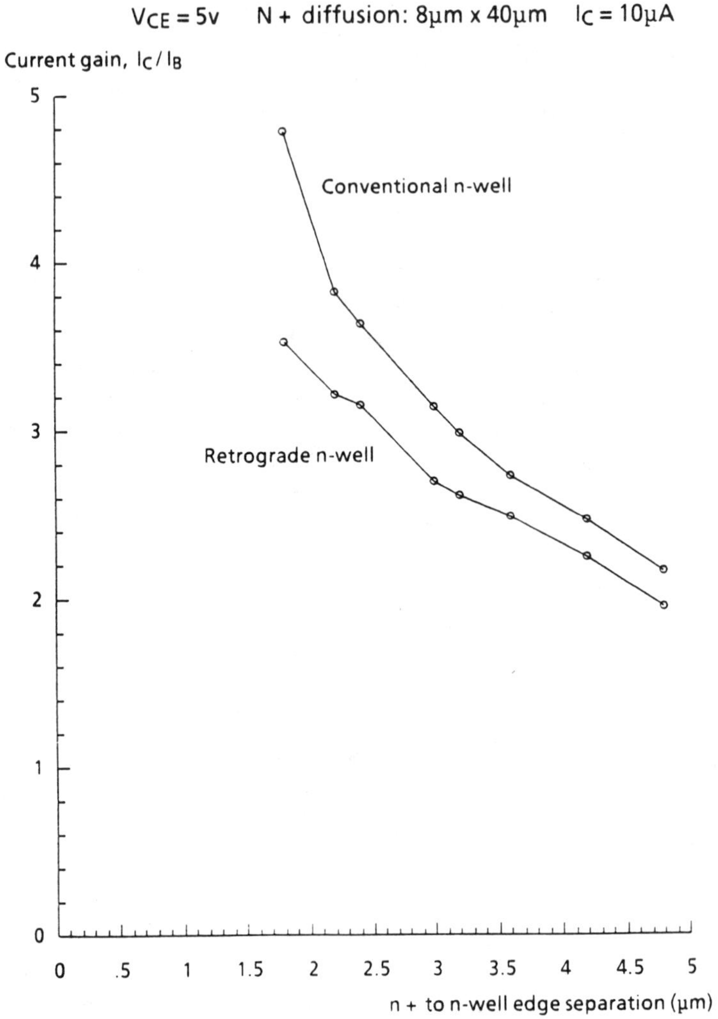

$V_{CE} = 5v$ N + diffusion: 8μm x 40μm $I_C = 10μA$

Figure 8.3   Lateral *npn* bipolar gains versus n^+ to *n*-well separation for conventional and retrograde *n*-wells (Ref. 6).

order of magnitude lower than that of a vertical bipolar device.   Fig. 8.3 shows the current gain of a lateral *npn* as a function of the n^+-to-*n*-well separation.[6]   The base transit time is also much larger due to the longer base width.   This weak lateral transistor not only reduces latchup susceptibility under steady state conditions, but also increases the response time necessary for causing transient latchup.   However, as the layout spacing decreases for high-density VLSI, this lateral bipolar device becomes stronger.

8.3.3 Parasitic Resistors

The two resistor values in the lumped circuit model can be estimated by SUPREM simulation or a spreading resistance measurement.   The well resistance (R_W) is basically the well sheet resistance under the diffusion inside the well multiplied by the number of squares in the layout.   Typical values are a few thousand or a few hundred ohms for conventional or retrograde n-well test structures.   The substrate resistance (R_S) is more complicated because of the spreading of the current in the substrate.   Two- or even three-dimensional numerical simulations and transmission line models have been developed and used to calculate the substrate resistance based on processing parameters and layout configuration.   The typical value is a few thousand ohms, but can drop to a few tens of ohms if n-on-n^+ epi is used.   The highly-doped substrate greatly reduces the R_S value.

8.4 LATCHUP CHARACTERIZATION

Latchup is often characterized by measuring the total current through the *pnpn* path while overstressing the anode voltage.[7]   Typical I-V characteristics for avalanche induced latchup are shown in Fig. 8.4.   Such an I-V curve is always seen when latchup is measured by a curve tracer.   Because a trace is

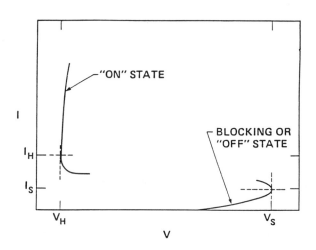

* V_S and I_S are the volatge and current at the switching point, V_H and I_H are the voltage and current at the holding point

Figure 8.4   Typical I-V characteristics of avalanche induced latchup.

made when the voltage is swept from the maximum voltage to zero as well
as from zero to the maximum voltage, the negative differential resistance
region is exhibited. The switching voltage V_S is the maximum voltage below
which the device is in a blocking or OFF state. At V_S, both bipolar devices
become active and, if sufficient loop gain is present, the *pnpn* regeneration
switches the device to a high-current, low-voltage state, referred to as the
ON or latched state. The current at this switching point, I_S, is also referred
to as the critical current for entering a latched state. The voltage required
to sustain this state, shown as V_H in the figure, is usually referred to as the
holding voltage. This voltage is the one at which the device enters into the
low-impedance state. The current at this voltage is called the holding current
(I_H).

8.4.1 Steady State Latchup Triggering

To trigger latchup, one of the two parasitic bipolar transistors must
become active. In an *n*-well configuration, this activation can be done by
forward biasing an n^+-to-*p*-substrate or a p^+-to-*n*-well junction. Two methods can be used to forward bias the two *pn* junctions. The first method is
to overstress one of the emitters by raising the p^+ voltage above V_{DD} or
lowering the n^+ voltage below V_{SS}. The second method is to debias one of
the bases by changing the *n*-well or the *p*-substrate bias. In the first method,
the entire voltage across the *pnpn* SCR is higher than V_{DD} whereas in the
second method the voltage drop remains at V_{DD}. The higher the voltage
across the SCR, the larger the current between the supply rails, hence latchup
is more likely to occur according to Eq. 8.5. Consequently, triggering by
overvoltage stress is more important because it represents the worst case.

Triggering by p^+ overvoltage. Figure 8.5(a) shows p^+ overvoltage

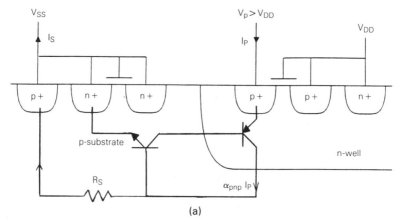

(a)

Figure 8.5 Latchup triggering by p^+ overvoltage: (a) measurement technique; (b)
I-V characteristics (Ref. 6).

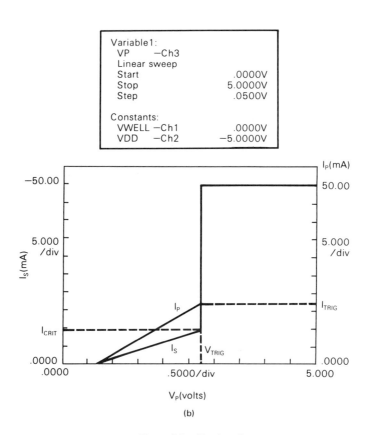

Figure 8.5 *Continued.*

triggering. The p^+ diffusion is ramped above V_{DD} while the supply is held at V_{DD}, which is 5 V. The current forced into the p^+ just before latchup is triggered is the trigger current, and the corresponding supply current is defined as the critical current. Figure 8.5(b) shows the I-V characteristics measured using a Hewlett Packard HP4145 parameter analyzer. As the voltage at the p^+ region (V_p) increases, both the current flowing to the p^+ region (I_p) and the supply current (I_s) increase until the currents are so large that latchup is induced, marked as the sharp rise in both currents. I_p and I_s just before this increase occurs are the trigger and critical current, respectively. They can be expressed as

$$I_p = I_{\text{TRIG}} \cong V_{ben}/(\alpha_{pnp}R_s) \tag{8.7a}$$

$$I_s = I_{\text{CRIT}} \cong V_{ben}/R_s \tag{8.7b}$$

Notice that I_p is larger than I_s before the triggering point because I_p is the emitter current and I_s is the collector of the pnp transistor and the npn is not yet turned on. α_{pnp} is the common base current gain for the vertical pnp transistor. As described in Chapter 5, α, defined as I_C/I_E, is less than one and is related to β as $(1 + 1/\beta)^{-1}$. V_p at the onset of triggering is defined as the trigger voltage.

Triggering by n^+ overvoltage. Figure 8.6 shows n^+ overvoltage triggering. The n^+ diffusion is ramped below V_{SS} while the V_{SS} is grounded. Again, the current forced into the n^+ node (I_n) just before latchup is triggered is the trigger current, and the corresponding supply current (I_d) is defined as the critical current. Simple expressions similar to Eq. 8.7 can be written as

$$I_n = I_{\text{TRIG}} \cong V_{bep}/(\alpha_{npn}R_W) \tag{8.8a}$$

$$I_d = I_{\text{CRIT}} \cong V_{bep}/R_W \tag{8.8b}$$

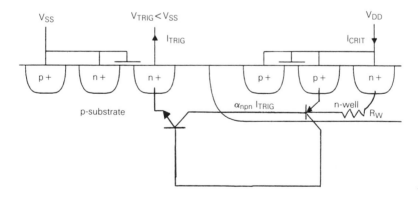

Figure 8.6 Latchup triggering by n^+ overvoltage, i.e., forward-bias n^+ to p-substrate junction.

Again, the trigger current is larger than the critical current because $\alpha_{npn} <$ 1, where α_{npn} is the common base current gain for the lateral *npn* transistor. Now the voltage at n^+ at the onset of triggering is defined as the trigger voltage.

Figure 8.7 shows the trigger and critical currents versus n^+-to-p^+ separation for both p^+ and n^+ overvoltage induced triggering.[8] The trigger current is always larger than the corresponding critical current as described. In the case of n^+ overvoltage stress, the trigger current increases rapidly as n^+-to-p^+ separation is enlarged.   This instance can be explained by referring

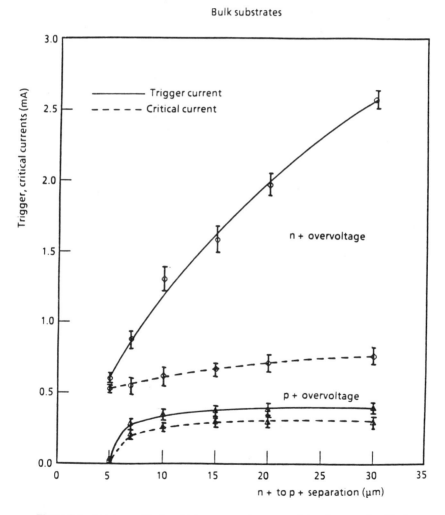

Figure 8.7   Latchup critical and trigger currents versus n^+ to p^+ separation (Ref. 8).

to Eq. 8.8(a), which shows that the trigger current is inversely proportional to α_{npn}, and this lateral bipolar gain is strongly dependent on the n^+-to-n-well spacing which is 40% of the n^+-to-p^+ separation in this example.   As shown in Fig. 8.3, β_{npn} decreases as n^+ to n-well spacing is increased; hence α_{npn} is expected to decrease in a similar rate as the n^+-to-p^+ separation is increased.

8.4.2 Steady State Latchup Holding

The supply current and voltage required to sustain latchup are defined as holding current and holding voltage.   To characterize them, a device is first triggered into the latched state by a method such as overvoltage stress, then the holding conditions are found by removing the triggering stimulus and ramping the supply voltage down until the device returns to the high impedance state.   Fig. 8.8(a) and (b) show the schematic and I-V characteristics in measuring holding conditions.   Resistors R_{S2} and R_{W2} are now added for more accurate analysis.   Latchup is first triggered by p^+ overvoltage, then V_{DD} is ramped down until the power supply current (I_D) suddenly drops.   The voltage and current at this point are denoted as the holding voltage and holding current.[9]   Modelling the holding condition is much more difficult than off state or triggering modelling[3] because at least one bipolar transistor is in saturation.   Forward bipolar gains characterized in a normal manner may not be applicable.   As shown in Fig. 8.8a, the total supply current at the holding condition is

$$I_D = I_{\text{HOLD}} = I_S + I_W + I_{BP} + I_{BN} \qquad (8.9a)$$

and

$$V_{DD} - V_{SS} = V_{\text{HOLD}} = V_{cep} + V_{ben} + (I_S + I_{BN})R_{S2} \qquad (8.9b)$$

Base current I_{BP} and I_{BN} in general are small.   Assuming I_{BN} can be neglected in Eq. 8.9(b), the holding voltage can be approximated as

$$V_{\text{HOLD}} \cong V_{cep} + V_{ben} + I_S R_{S2} \qquad (8.9c)$$

$$= V_{cep} + V_{ben} (1 + R_{S2}/R_{S1}) \qquad (8.9d)$$

where V_{cep} is the collector emitter voltage of the *pnp* transistor.   The holding voltage provides a convenient way to calibrate latchup immunity.   If the holding voltage is maintained above the supply voltage, then no permanent latchup state can be sustained.   This condition is, however, difficult to achieve in practice and relaxed latchup resistance is often tolerated.

According to Eq. (8.9d), the holding voltage can be raised by increasing R_{S2} through the increase of n^+-to-p^+ separation, or decreasing R_{S1} with decreasing epi thickness.   However, the resistors in the lumped circuit model are distributed and cannot be modelled accurately by discrete elements.   More

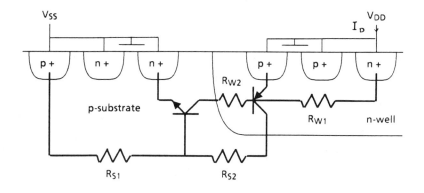

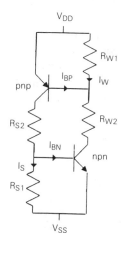

(a)

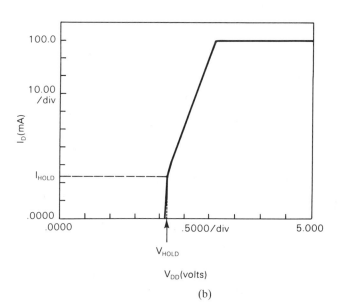

(b)

Figure 8.8 (a) First order lumped equivalent circuit model for latchup; and (b) I-V characteristics during holding condition (Ref. 6).

sophisticated models such as a transmission line model[10] or 2-D numerical simulators[11,12] should be used to make better predictions.

8.4.3 Transient Latchup

Although steady-state characterization represents the worst case for latchup susceptibility, transient testing provides an understanding of latchup behavior at various conditions which may only occur during transient upsets in actual circuit operations.

Figure 8.9(a) and (b) show the testing arrangements for obtaining transient latchup characteristics. Overvoltage stress is applied by a pulse rather than a constant bias on the p^+ or n^+ diffusion. For a given pulse height, the pulse width is increased until latchup is triggered.

Figure 8.10 shows the minimum pulse width needed for triggering latchup at various pulse heights applied to the p^+ diffusion.[6] At large pulse widths, pulse height approaches the steady state triggering condition. As the pulse

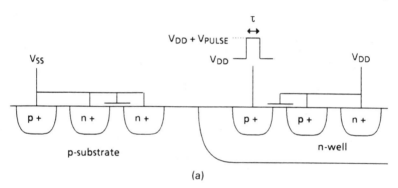

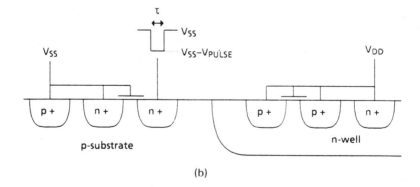

Figure 8.9 Transient latchup measurement setups: (a) p^+ overvoltage triggering; and (b) n^+ overvoltage triggering (Ref. 6).

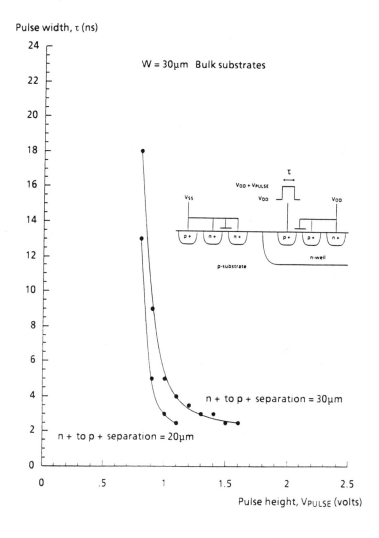

Figure 8.10 Pulse width required to trigger latchup for various pulse heights (Ref. 6).

width is decreased and becomes comparable with bipolar response time, the pulse height required to trigger latchup increases, indicating that latchup is more difficult to induce. When the pulse width is further reduced, eventually latchup cannot be triggered regardless of pulse height. The bipolar devices simply cannot respond quickly enough to enter the latched state before the triggering stimulus is removed. The minimum pulse width τ_{min} for triggering latchup may be approximated by the sum of the vertical and lateral bipolar base transit times[13]:

$$t_{min} = \tau_{pnp} + \tau_{npn} = \tau_{pnp} + W_n^2/\eta D_n \qquad (8.10)$$

where W_n is the lateral npn basewidth, corresponding to the n^+-to-n-well spacing. It is therefore expected that the curve in Fig. 8.10 will shift downward as the n^+-to-p^+ separation is reduced.

8.4.4 Field Inversion Effects

Field inversion can initiate latchup. Surface leakage induced by parasitic FETs has been reported[14] to initiate latchup at a lower V_{DD} voltage, defined as the critical voltage (V_{CRIT}). This phenomenon is illustrated by the p-well CMOS structure as shown in Fig. 8.11. Notice that the two parasitic FETs are p^+-to-p-well and n^+-to-n-substrate FETs. If the field oxide gate voltage (V_{GS}) reaches V_{TFN}, threshold of the n^+-to-n-substrate parasitic FET, electron current injects into the n-substrate, thus forward biasing the pnp bipolar. The npn is already on when a FET current is present because the n^+/p junction is forward-biased at the Si surface near the gate. Therefore, the V_{CRIT} needed for latchup decreases sharply as $V_{GS} > V_{TFN}$. For $V_{GS} < V_{TFN}$, $V_{GS} - V_{CRIT}$ is less than V_{TFP}, the threshold of the p^+-to-p-well parasitic FET, thus turning the p-channel parasitic FET on. The hole current injects into the p-well, forward-biasing the npn bipolar. Again, because the pnp is already on, latchup is triggered by the hole current initiated at $V_{GS} -$

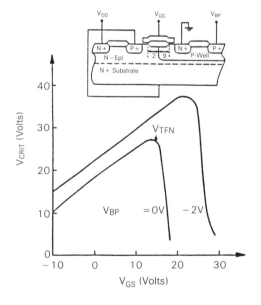

Figure 8.11 Field inversion induced latchup. The critical voltage (V_{CRIT}) is the V_{DD} at which the device is switched into latchup. V_{CRIT} is a function of field electrode voltage (V_{GS}) for two p-well biases (Ref. 14, © 1982 IEEE).

$V_{\text{CRIT}} = V_{TFP}$. So, for -10 V $< V_{GS} < V_{TFN}$, V_{CRIT} must increase as V_{GS} is increased to turn on the p-channel parasitic FET. As shown in the figure, V_{TFN} increases as the substrate voltage (V_{BP}) is applied due to the well-known body effect. Moreover, when the p-well is biased at a negative voltage (V_{BP}) and the n^+ in the well is at ground, V_{CRIT} must be raised to supply more hole current to forward bias the n^+-to-p-well junction. Consequently, higher V_{CRIT} values are observed in the case of $V_{BP} = -2$ V.

8.5 AVOIDING LATCHUP

Based on the lumped circuit model expressed in Eq. 8.5, it is clear that latchup can be prevented by either reducing the bipolar gains or lowering the shunting resistances. Numerous methods have been proposed to eliminate or reduce latchup possibility. However, all of them can be categorized into two classes: bipolar spoiling and bipolar decoupling.[15]

8.5.1 Bipolar Spoiling

In the case of bipolar spoiling, lifetime reduction by gold doping[16] and neutron radiation[17] have been tried. However, this approach introduces leakage and is difficult to control. Other methods such as the use of Schottky barrier source/drains[18,19] have also been employed to decrease emitter efficiency, but often at the expense of degrading FET performance.

If the gain product ($\beta_{npn}\beta_{pnp}$) can be kept below one, then no latchup is possible; however, such a condition cannot usually be met in practice, as previously discussed, particularly as lateral and vertical circuit dimensions are reduced. Moreover, FET performance often suffers when processing techniques are introduced to damage bipolar devices.

As previously mentioned, retrograde well technology has been an effective way to reduce the vertical bipolar gain by providing a high Gummel number[20] and a retarded electric field. An order of magnitude reduction in the vertical bipolar gain can easily be achieved. In addition, the lower resistance associated with the retrograde well is helpful in providing latchup decoupling.

8.5.2 Bipolar Decoupling

In addition to the use of a retrograde well to reduce well resistance, other bipolar decoupling methods include using epitaxial materials on heavily-doped substrates,[21] guard rings[1] and dielectric isolation.[22] Because the use of guard rings requires additional layout areas, they have often been used only for I/O circuits, so that layout penalty is minimal. Other techniques do not need extra layout areas and can be applied to the entire circuit.

Retrograde well.   A retrograde well provides a higher doping concentration at the bottom of the well, hence a low well resistance (R_W) can be obtained.   The low R_W offers an effective path in shunting the well current, decreasing the coupling of the two bipolar devices.   Improvement is most significant when latchup is triggered by n^+ overvoltage.   As described by Eq. 8.8, both trigger and critical currents during n^+ overvoltage stress are inversely proportional to R_W, the well resistance.   With the retrograde-well approach, R_W can be reduced by one order of magnitude, which results in about a factor of ten increase in trigger current.   Experimental results obtained by comparing a retrograde and conventional well are shown in Fig. 8.12.[6]   Moreover, the retrograde well approach can improve latchup resistance indirectly because it allows the use of a thinner epitaxial layer, which is very effective in latchup protection.

Epitaxial substrate.   As described in Sec. 8.2, the well and substrate provide two parasitic resistors (R_W and R_S), which sink a portion of the transistor collector currents to the power supply rails (V_{SS} and V_{DD}), thereby reducing positive feedback induced by bipolar coupling.   This decoupling of the bipolar devices becomes more effective as R_S and/or R_W are reduced. R_W can be reduced by increasing well doping concentration; but a too heavily-doped well can result in high junction capacitance and large body effect for the active FET devices in the well. R_S can be decreased by using an epitaxial layer on a heavily doped substrate.   Because the heavily-doped substrate is farther away from the active channel region, this method, to first order, does not affect the FET performance.   As a result, epitaxy has become an effective means of latchup decoupling.[3]   The thinner the epi layer, the smaller the value of R_s and the more latchup-resistant the circuit becomes.   Not only does epi result in a reduction of R_s, it also offers a quasi-ground plane that sinks a larger portion of the vertical bipolar collector current into the substrate.   With a thin epi layer, the quasi-ground plane offered by the heavily-doped substrate is near the bottom of the well.   The ground plane can then more easily sink the collector current of the vertical bipolar into the substrate. In general, latchup immunity is improved with a thin epi layer at either a triggering or holding condition.

Figure 8.13 shows the trigger current and holding voltage as a function of n^+-to-p^+ spacing for various epi thicknesses.[23,24]   The epi thicknesses specified in the figure are the effective (or final) epi thicknesses.   Because of out-diffusion of impurity atoms from the heavily-doped substrate, the effective epi thickness after wafer processing is thinner than the initial as-grown epi thickness.   Out-diffusion arises from two major sources: high temperature epi growth and thermal diffusion during well formation.   The effective epi thickness here is defined as the thickness of the epi layer with a doping concentration less than 1×10^{17} cm^{-3}.   Very thin epi can only be realized with a retrograde well, which is formed by a high energy implant rather than

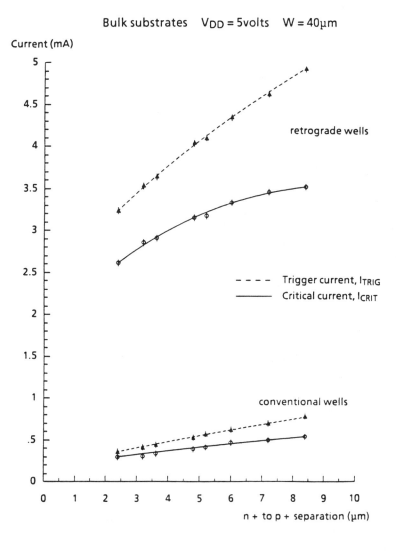

Figure 8.12 Latchup trigger and critical currents for conventional and retrograde *n*-wells (Ref. 6).

diffusion.   It is clear from this figure that thin epi improves latchup resistance by almost one order of magnitude from either a triggering or holding point of view.   The advantage however diminishes rapidly as the n^+-to-p^+ separation becomes comparable to the epi thickness.

Scaling epi thickness with n^+-to-p^+ separation.   A systematic scheme has been proposed[12] to maintain latchup immunity in scaling CMOS to micron

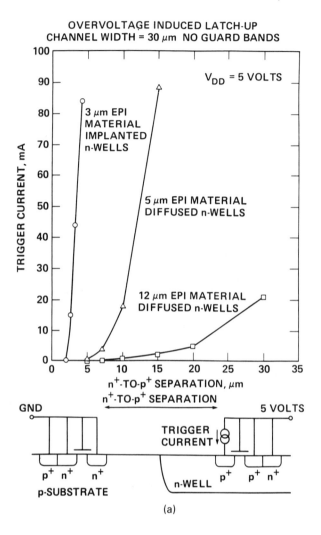

Figure 8.13 (a) Latchup trigger current; and (b) holding voltages versus n^+ to p^+ separation for epi substrates with different epi thicknesses (Ref. 23, © 1986 IEEE).

and submicron technologies. Figure 8.14 shows the critical currents plotted against the n^+-to-p^+ spacing normalized by the minimum feature size.[12] Measurement setup is inserted in the figure. Minimum feature sizes are 1.2 μm and 2 μm for the 1.2 μm and 2 μm processes, respectively. For the 2 μm process, notice that the critical current, measured when the p^+ is triggered, is much higher for the thinner epi case, indicating latchup is more

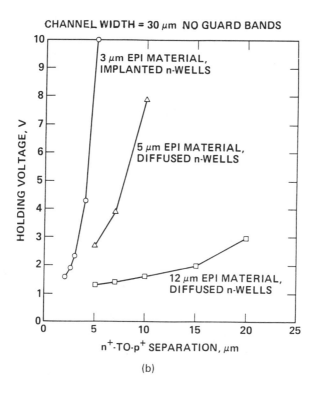

CHANNEL WIDTH = 30 μm NO GUARD BANDS

(b)

Figure 8.13 *Continued.*

difficult to induce.   In addition, the results for the 1.2 μm process on 12 μm epi is very similar to those for the 2 μm process on 20 μm epi.   This similarity means that the critical current during latchup triggering can be maintained if the epi thickness is scaled with the feature size.

Latchup scaling also works well for holding voltage.   In this scheme, a critical separation between n^+ and p^+ is defined, which ensures a holding voltage at 5 V, the present power supply voltage.   If the holding voltage is above 5 V, latchup cannot be maintained with the standard 5 V supply, and the circuit is latchup immune.   Simulated and measured results from various CMOS technologies show that the critical n^+-to-p^+ separation scales linearly with epitaxial layer thickness as shown in Fig. 8.15.[12]   The simulated results were obtained using a numerical model that solves the current continuity equation, $\nabla \cdot [\sigma(x,y)\nabla\Phi(x,y)] = 0$, in two dimensions.[12]   Specifically, latchup immunity can be maintained if the final epi thickness is about 2/3 of the n^+-to-p^+ layout spacing.   The actual doping profile in the well (conventional or retrograde well) does not significantly affect this result.   To act as a ground plane, the substrate doping concentration however should be at least two

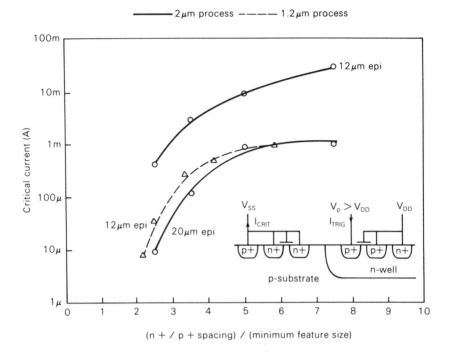

Figure 8.14 Latchup critical currents versus normalized n^+ to p^+ separation for two processes with two final epi thicknesses. In both cases, final epi thickness is 10 times the minimum feature size (Ref. 12, © 1986 IEEE).

orders of magnitude greater than that in the epitaxial layer. A doping level at 10^{17} cm^{-3} or above should be used for the substrate.

Limitation in epi thinning.   It has already been pointed out that p-on-p^+ epitaxial substrate material is very effective at suppressing latchup. However, as the n^+-to-p^+ separation is reduced, thinner and thinner epitaxial layers are required. This requirement presents a problem because of the out-diffusion of boron from the substrate during processing. The transition from the heavily doped substrate to the lightly doped epitaxial layer occurs over several microns, limiting the minimum epitaxial layer thickness that can be used. The greatest amount of out-diffusion occurs during the n-well drive-in. By so eliminating this step, as in the case of retrograde well technology, thinner epitaxial material can be used. In practice, the thinnest epitaxial material that can be used with the conventional well process used in this example is about 4 μm, while layers as thin as 2 μm can be used with retrograde well technology.

Figure 8.16 shows measured latchup holding voltages as a function of n^+-to-p^+ separation for devices on thin epitaxial material.[6] Results are shown

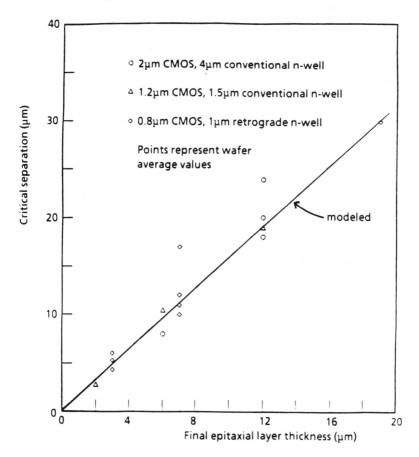

Figure 8.15   Critical n^+ to p^+ spacing for 5 V latchup holding voltage (Ref. 12, © 1986 IEEE).

for both retrograde and conventional n-well devices fabricated on material with a 4 μm starting epitaxial layer thickness, and for retrograde well devices only on material with a 2 μm starting epi thickness.

Comparing the curves for the 4 μm (as grown) material in Fig. 8.16, conventional structures appear to have better performance in comparison to retrograde well devices. The final (or effective) epi layer of conventional devices is thinner due to the greater out-diffusion of boron from the heavily doped p^+ substrate during the well drive-in. Within conventional wells, increasing the n^+ to p^+ separation to about 4 μm gives a latchup holding voltage of more than 5 volts, and immunity from permanent latched states. Because the thermal budget is reduced for retrograde-well devices, the corresponding final (effective) epi layer is thicker than that of conventional well devices,

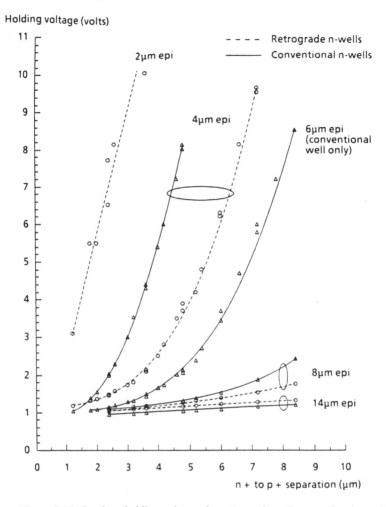

Figure 8.16 Latchup holding voltages for retrograde and conventional *n*-wells made in epi substrates with several as-grown epi thicknesses (Ref. 6, © 1987 IEEE). The corresponding test structure and measuring technique are shown in Fig. 8.8.

the equivalent separation for these devices on the same material is about 6 μm.

A better comparison is between devices formed on the thinnest epitaxial material allowed by the technology type, that is, retrograde well devices on 2 μm epitaxial material and conventional devices on 4 μm material. The greater latchup immunity achieved using retrograde structures is then clear, with the holding voltage reaching 5 volts at only 1.8 μm n^+-to-p^+ separation. This dimension happens to coincide with the minimum separation needed for adequate isolation between opposite-type devices (see Chapter 7).

Trench isolation.   The trench isolation technique and associated problems were discussed in Chapter 7.   Trench or other dielectric isolation such as SEG (Selective Epi Growth) provides a physical decoupling between an n- and a p-channel device, thus enhancing latchup protection.   The use of deep trenches on epi substrate has shown a latchup-free CMOS structure.[25] The lateral current to and from the well is greatly reduced, if not completely eliminated.   However, trench isolation alone does not significantly increase trigger or holding current.[26]   The major advantage of trench isolation alone is the increase of latchup response time, resulting in substantial improvement in preventing transient upset.

8.6 LAYOUT CONSIDERATIONS

Latchup can also be avoided by modifying circuit layout.   Two major layout considerations are the placement of well and substrate contacts and the inclusion of guard bands.   Width effects should also be taken into account.

8.6.1 Well and Substrate Contacts

To avoid latchup, the well and substrate resistances should be minimized for effective bipolar decoupling.   The resistance value is the product of the sheet resistance (Ω per square) and the resistor length in terms of the number of squares.   As described in the previous section, the well or substrate sheet resistance can be reduced by using a retrograde well or an epitaxy layer, respectively.   The resistor length can be decreased by shortening the distance between the well or substrate contact to the FET; that is minimizing the distances d_n and d_p shown in Fig. 8.17.

Triggering.   With n^+ overvoltage triggering, the npn emitter is already forward-biased, so the critical current for latchup to occur is relatively independent of the location of the substrate contact.   As expressed in Eq.

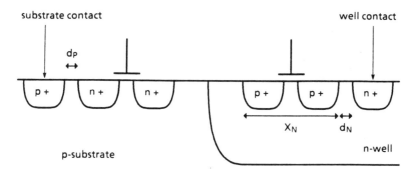

Figure 8.17 CMOS inverter structure showing well and substrate contacts.

8.8(b), the critical current only depends on the ohmic drop at the *pnp* emitter-base junction and the well resistance, which can be approximated by

$$R_W = \rho_w \cdot (X_N + d_N)/W \qquad (8.11)$$

where X_N is defined in Fig. 8.17, W is the FET width, and ρ_w is the well sheet resistance. Substituting Eq. 8.11 into Eq. 8.8(b) results in

$$I_{\text{CRIT}} = V_{bep}W/[\rho_w \cdot (X_N + d_N)] \qquad (8.12)$$

Figure 8.18 shows the dependence of critical current on d_P and d_N when latchup is triggered by n^+ overvoltage stress.[8] The dashed line represents the modelled results using a value of 2500 $\Omega/\square$ for ρ_w. As the well contact is placed closer to the pMOSFET source, the critical current increases, suggesting an improvement in latchup immunity. It is also helpful to place multiple contacts so that each FET always has a well contact in its proximity. Notice that the critical current is relatively independent of d_P, meaning the layout of the substrate contact is not critical in this case.

   With p^+ overvoltage triggering, the *pnp* emitter is already forward-biased, so the critical current for latchup to occur is relatively independent of the location of the well contact. The critical current should depend on

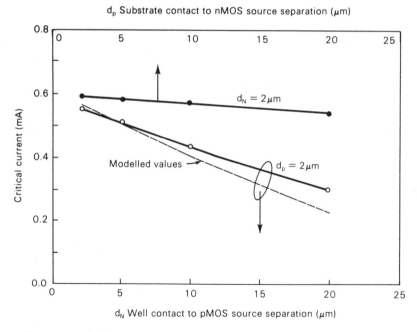

Figure 8.18 Critical currents at n^+ overvoltage triggering versus the separation between well (or substrate) contact to *p*MOS (or *n*MOS) source. (After Lewis, Ref. 8.)

the substrate contact location.   The dependence, however, is generally very weak because the *pnp* collector current is also collected at the back substrate contact.   In essence, the placement of the substrate contact in layout is less important if a backside contact is provided.

Holding.   The effect of layout on the latchup holding condition has also been examined[27] using the three test structures A, B and C shown in Fig. 8.19(a), (b), and (c), respectively.   The well (or substrate) contact to the corresponding FET source is 10 μm in structure A, 3 μm in structure B, and 0 μm in structure C, because butting contacts are used there.   The holding voltage and holding current as a function of n^+ to p^+ separation (d) are shown in Fig. 8.19(d) for the three structures.[27]   Both holding current and

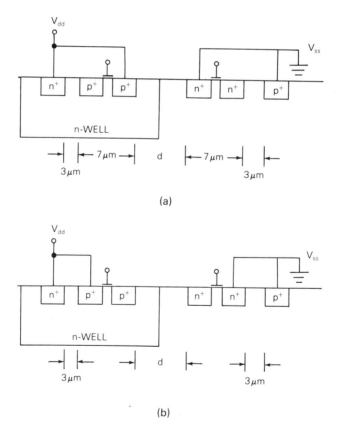

(a)

(b)

Figure 8.19   Cross-sections of latchup test structures: (a) source and contact separated by 10 μm; (b) source and contact separated by 3 μm; (c) butted source and contact; and (d) holding voltage and holding current as a function of n^+ to p^+ separation (Ref. 27, © 1984 IEEE).

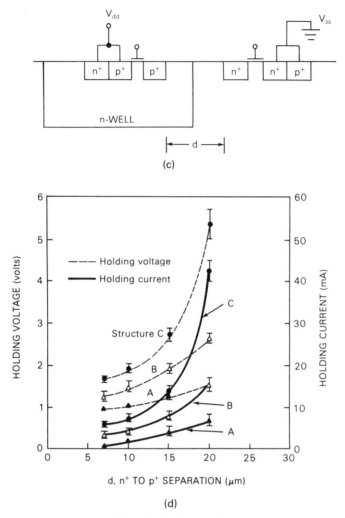

Figure 8.19 *Continued.*

holding voltage increase with an increase of n^+-to-p^+ separation. For any given n^+-to-p^+ separation, both the holding voltage and holding current increase as the well and substrate contacts are moved closer to, and eventually abut to, the FET sources.

8.6.2 Guard Bands

The n^+ or p^+ diffusion used for the well or substrate contact can be placed between the n- and p-channel FETs, as shown in Fig. 8.20. In this configuration, majority carriers are pre-collected by the contact diffusions

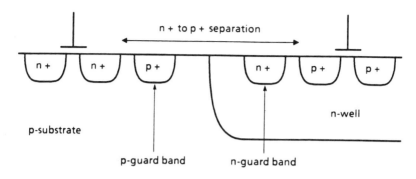

Figure 8.20   CMOS inverter structure showing guard bands.

before they inject to a bipolar base or cause an ohmic drop in the well or substrate. These diffusions are referred to as guard bands, which are commonly used in input/output circuits, where latchup is more of a problem because larger currents often exist and extra current sometimes injects into the circuit from the I/O pads. Another reason is that for just I/O circuits, the penalty due to additional layout area is not great. Figure 8.21(a) shows the n^+ guard band in an n-well.[3] As shown in the figure, the n^+ guard band serves the purpose of current steering, and it becomes even more effective

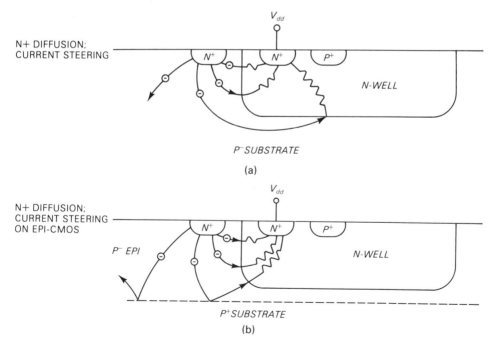

Figure 8.21   Majority carrier guard (N^+) in well to steer current away from vertical pnp: (a) p-substrate; (b) p-epi on p^+ substrate. (After Troutman, Ref. 3.)

when epi is used. This increased effectiveness is due to the built-in field at the p^-/p^+ junction. It tends to reflect the electrons injecting downward so that they can be collected by the guard band (Fig. 8.21b).

Similar effects appear for the p^+ guard band in a p-substrate (Fig. 8.22a). The p^+ guard band steers the hole current away from the n^+ junction. The effect of epi is, however, different. Because the p^-/p^+ junction and the highly-conductive p^+ substrate tend to absorb the most current, it is preferable to have a substrate contact p^+ guard ring, which can further reduce any voltage drop near the n^+ junction.

The effects of guard bands on latchup triggering are shown in Fig. 8.23.[28] In those experiments, transient triggering was done by debasing the substrate or well to forward-bias the npn or pnp emitter/base junction. The total power supply across the $pnpn$ parasitic diode structure was maintained at V_{DD}. As expected, the pulse height needed to trigger latchup decreased and approached the steady state value as the pulse width increased. For the npn triggering shown in Fig. 8.23(a), the n^+ guard band is effective in improving latchup resistance. For the pnp triggering shown in Fig. 8.23(b), the use of a p^+ guard band can improve latchup immunity. It is clear that improvement is greatest when both n^+ and p^+ guard bands are used.

n^+ can also be placed in the p-substrate to collect electrons before they inject into the well.[3] This n^+ guard band serves as a minority carrier collector

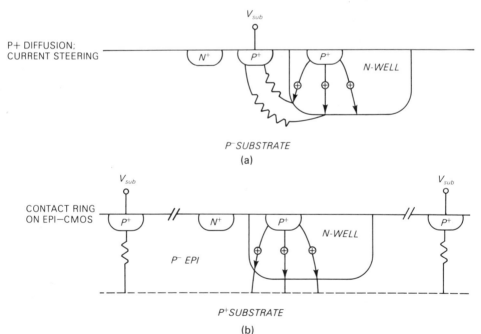

Figure 8.22 Majority carrier guard (P^+) in substrate to steer current away from lateral npn: (a) p-substrate; and (b) p-epi on p^+ substrate. (After Troutman, Ref. 3.)

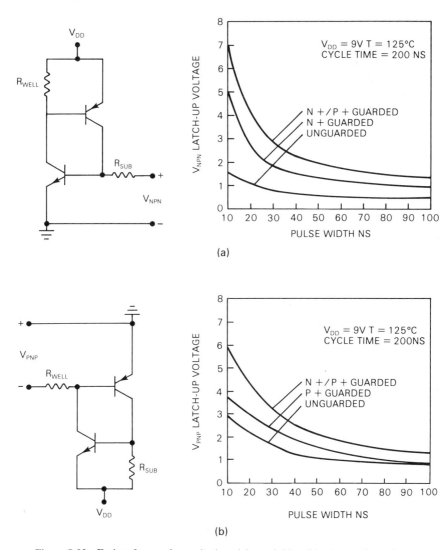

Figure 8.23 Emitter/base pulse excitation. (a) *npn* initiated latchup and npn forward-bias vs. pulse width; and (b) *pnp* initiated latchup and *pnp* forward-bias vs. pulse width for unguarded and guarded structures (Ref. 28, © 1983 IEEE).

(Fig. 8.24) rather than a majority carrier collector, as previously described. An *n*-well, which is much deeper than the n^+, is often used under the n^+ to increase collection efficiency (Fig. 8.24b).   The current collected by this *n*-well guard band (I_{C1}) is shown in Fig. 8.25(b) and is much larger than the current injected into the real *n*-well (I_{C2}).[29] V_E is the n^+/P junction bias, as shown in the measurement schematic (Fig. 8.25(a)).   Again, a guard band

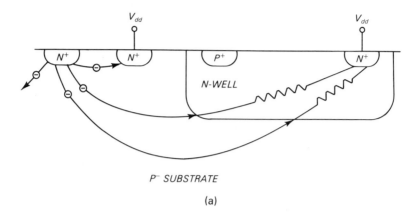

(a)

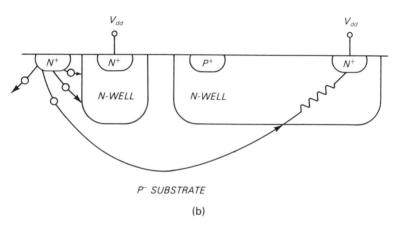

(b)

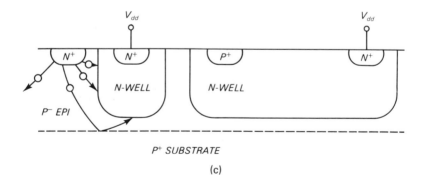

(c)

Figure 8.24 Minority carrier guard in substrate: (a) N^+ diffusion guard; (b) N-well guard; and (c) N-well on epi. (After Troutman, Ref. 3.)

is more effective when epi is used (Fig. 8.24c). As shown in Fig. 8.25(c), the current escaping from the guard collection and injected to the actual n-well (I_{C2}) is greatly reduced.

8.6.3 Three-dimensional Effects

Three-dimensional layout effects on latchup have been reported.[30]   It was found that narrow width effects were significant in determining latchup immunity, and resulted in a significant deviation from ideal scaling width. This deviation is due to the three-dimensional current flow and has been modelled numerically by a 3-D current continuity equation, i.e., $\nabla \cdot (\sigma \nabla \Phi(x,y,z)) = 0$. Figure 8.26 shows calculated surface potential contours for two devices with different diffusion widths.[30]   Notice that the potential and current flow at the corners of the p^+ emitter are really three-dimensional. For the narrow device, the equipotential contours do not extend as far as the wide device due to this corner effect. The potential at the middle of the p^+-to-n-well gap (marked X in the figure) is 0.7 V for the wide device, but $\leq$ 0.5 V for the narrow device. If the n^+ source is located nearby, the wide device may be latched, but not the narrow one.

Measurements from devices with various widths show that both triggering and holding conditions are dependent on device width. Holding voltages for different epi thicknesses are shown in Fig. 8.27.[30]   In general, holding voltage increases as diffusion width is reduced. Latchup immunity for narrow-width devices may be significantly better than what is predicted by 2-D modelling. Conversely, latchup test structures are often not very wide to limit current drawn in the latched state, and the latchup susceptibility of wider devices in a real circuit layout may be worse. These three-dimensional effects should not be ignored in actual circuit layout.

8.7 SUMMARY

In scaling CMOS for VLSI, latchup has become a major problem due to the tight separation between an n- and a p-channel FET. Latchup is normally characterized by triggering and holding conditions. The triggering condition refers to the current and voltage needed to initiate latchup while the holding condition corresponds to the voltage and current required to sustain a latched state once it is triggered. The physics and condition for latchup to occur have been described using a lumped circuit equivalent model. If the holding voltage is above the power supply voltage (5 V, in general), latchup cannot be sustained. Although current spreading in a three-dimensional manner is difficult to model using discrete components, the lumped circuit model provides a good basis for seeking solutions to the problem. Based on the model,

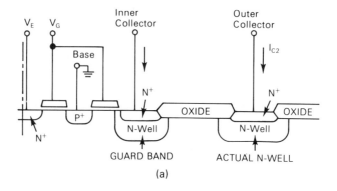

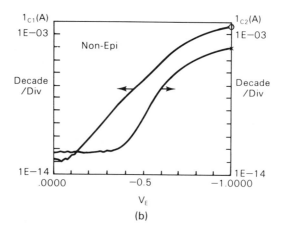

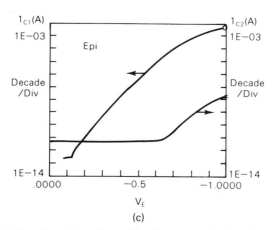

Figure 8.25 The effect of *n*-well guard band on current injected into the real *n*-well: (a) measurement setup; (b) I-V for non-epi; and (c) I-V for epi substrate. (After Troutman, Ref. 29.)

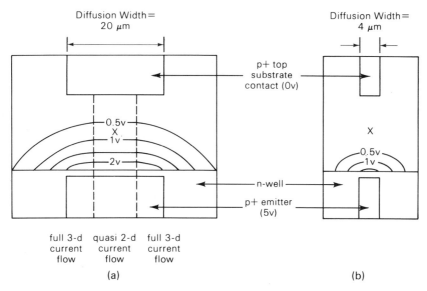

Figure 8.26 Surface potential contours calculated using 3-D current flow model. Epi thickness is 6 μm. n^+ S/D diffusion is marked by x, lying between p^+ contact and n-well. Diffusion widths are (a) 20 μm, (b) 4 μm. (Ref. 30, © 1986 IEEE).

it is easy to see that bipolar spoiling and bipolar decoupling are the two basic methods to avoid latchup.

While spoiling bipolar transistors in a VLSI circuit often has the risk of damaging active CMOS FETs, bipolar decoupling has been more effective in preventing latchup. The most common way to avoid latchup is to use a lightly doped epi layer on a heavily doped substrate, which acts as a ground plane to shunt vertical bipolar current to the substrate so that positive feedback can be avoided. The thinner the epi layer, the more effective it is in steering the current to the substrate. In fact, latchup immunity can be maintained when epi thickness is scaled with n^+-to-p^+ separation. However, dopant out-diffusion during epitaxial growth and subsequent well diffusion place a limit on minimum epitaxy thickness. Low temperature epitaxy and retrograde wells formed by a high-energy implant instead of diffusion allow the use of very thin (2–3 μm) epi layers for latchup immunity.[6]

Careful layout, especially of I/O circuits, should be exercised to offer latchup protection. Substrate and well contacts must be laid out near the FET sources and multiple well contacts and a backside substrate contact are highly recommended. Guard bands between n- and p-channel FETs are effective in steering current away from the FET sources, thereby reducing the likelihood of forward-biasing emitter junctions. Finally, narrow-width effects should be taken into account in determining latchup sensitivity in real CMOS circuits.

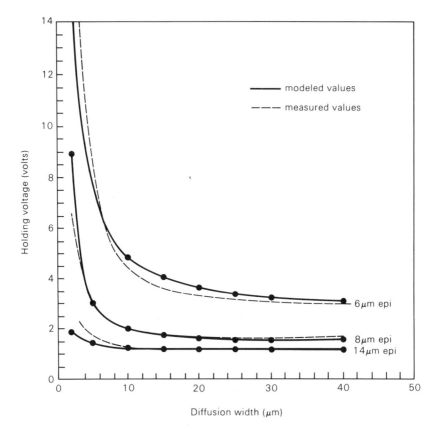

Figure 8.27 Latchup holding voltage as a function of diffusion width for various epi thicknesses (Ref. 30, © 1986 IEEE).

REFERENCES

1. D. B. Estreich, "The Physics and Modeling of Latch-Up and CMOS Integrated Circuits," Stanford Electronics Lab. Tech. Rep. G-201-9, 1980.

2. D. B. Estreich and R. W. Dutton, "Modelling latch-up in CMOS integrated circuits," *IEEE Trans. on Computer-Aided Design of Integrated Circuits and Systems*, vol. CAD-1, p. 157, 1982.

3. R. R. Troutman, *Latch-up in CMOS Technology: The Problem and Its Cure*, Kluwer Academic Publishers, 1986.

4. S. Sze, *Physics of Semiconductor Devices*, 2nd ed., John Wiley & Sons, 1981.

5. A. Grove, *Physics and Technology of Semiconductor Devices*, John Wiley & Sons, 1967.

6. A. G. Lewis, R. A. Martin and J. Y. Chen, "Retrograde and conventional *n*-well CMOS technologies: A comparison," *Internal Report*, Xerox Palo Alto Research Center, EIL-86-11, 1986; also in A. G. Lewis, R. A. Martin, T. Y. Huang,

J. Y. Chen, "Latchup performance of retrograde and conventional *n*-well CMOS technologies," *IEEE Trans. Electron Devices*, vol. ED-34, p. 2156, 1987.

7. R. R. Troutman, "Recent developments in CMOS latch-up," in *IEDM Tech. Dig.*, p. 296, 1984.

8. A. G. Lewis, "Latchup in the CMOS3 2 micron technology," *Internal Report*, Xerox Palo Alto Research Center, ICL-86-05, 1986.

9. A. G. Lewis, "Latchup suppression in fine-dimension shallow *p*-well CMOS circuits," *IEEE Trans. Elec. Dev.*, vol. ED-31, p. 1472, 1984.

10. R. R. Troutman and M. J. Hargrove, "Transmission line modeling of substrate resistance and CMOS latchup," *IEEE Trans. Elec. Dev.*, ED-33, p. 945, 1986.

11. G. J. Hu, "A better understanding of CMOS latchup," *IEEE Trans. Elec. Dev.*, ED-31, p. 62, 1984.

12. A. G. Lewis, R. A. Martin, T. Y. Huang, J. Y. Chen and R. H. Bruce, "Scaling CMOS Technologies with Constant Latch-Up Immunity," *Tech. Digest of 1986 Symp. on VLSI Tech.*, San Diego, CA , p. 23, May 1986.

13. R. D. Rung and H. Momose, "DC holding and dynamic triggering characteristics of bulk CMOS latch-up," *IEEE Trans. on Electron Devices*, vol. ED-30, p. 1647, 1983.

14. D. Takacs, C. Werner, J. Harter and U. Schwabe, "Surface Induced Latch-Up in VLSI CMOS Circuits," *IEDM Tech. Digest*, p. 458, 1982.

15. R. R. Troutman, "Recent development and future trends in latch-up prevention in scaled CMOS," 1983 Devices Research Conf., *IEEE Trans. on Electron Devices*, vol. ED-30, p. 1564, 1983.

16. W. R. Dawes, Jr. and G. F. Derbenwick, "Prevention of CMOS latch-up by gold doping," *IEEE Trans. Nucl. Sci.*, vol. NS-23, p. 2027, 1976.

17. J. R. Adams and R. J. Sokel, "Neutron irradiation for prevention of latch-up in MOS integrated circuits," presented at the 1979 Nuclear and Space Radiation Effects Conf. (Santa Cruz, CA), July 19, 1979.

18. M. E. Rebeschini and M. D. Sugino, "Latch-up-free CMOS by Schottky barrier technology," in the *Digest of Sym. on VLSI Technology*, p. 6, 1982.

19. S. Swirhun, E. Sangiorgi, A. Weeks, R. M. Swanson, K. C. Saraswat, and R. W. Dutton, "Latch-up-free CMOS using guarded Schottky barrier PMOS," in *IEDM Tech. Dig.*, p. 402, 1984.

20. H. K. Gummel, "Measurement of the number of impurities in the base layer of a transistor," *Proc. IRE*, vol. 49, p. 834, 1961.

21. D. B. Estreich, A. Ochoa, Jr., and R. W. Dutton, "An analysis of latch-up prevention in CMOS ICs using an epitaxial-buried layer process," in *IEDM Tech. Dig.*, p. 230, 1978.

22. D. Williams, C. Fa, G. Larchian, and O. Maxwell, "A new dielectric isolation technique for high-quality silicon integrated circuits," *Electrochem. Soc.*, vol. 111, p. 153.

23. A. G. Lewis, R. H. Bruce, W. G. Gunning and T-Y. Huang, "Latch-up induced in *n*-well CMOS circuits by signal clamping," *European Solid State Device Research Conf.*, Aachen, Germany, Sept. 1985; Also R. A. Martin, A. G. Lewis, T. Y. Huang, and J. Y. Chen, "A new process for one micron and finer CMOS," *IEDM Tech. Dig.*, p. 403, 1985.

24. J. Y. Chen, "CMOS—The emerging VLSI technology," *IEEE Circuits and Devices Magazine*," vol. 2, p. 16, 1986.

25. T. Yamaguchi, S. Morimoto, G. Kawamoto, H. K. Park, and G. C. Eiden, "High-speed latch-up-free 0.5 μm-channel CMOS using self-aligned Ti-Si and deep-trench isolation technologies, in *IEDM Tech. Dig.*, p. 522, 1983.

26. R. D. Rung, "Trench Isolation Perspects for Application in CMOS VLSI," in *IEDM Tech. Dig.*, p. 574, 1984.

27. G. J. Hu and R. H. Bruce, "A CMOS structure with high latch-up holding voltage," *IEEE Ele. Dev. Lett.*, EDL-5, p. 211, 1984.

28. E. Hamdy and A. Mohsen, "Characterization and modelling of transient latch-up in CMOS technology," in *IEDM Tech. Dig.*, p. 172, 1983.

29. R. R. Troutman, "Latch-up in CMOS," CMOS Short Course, UCLA, June 1984.

30. A. G. Lewis, R. A. Martin, T. Y. Huang, and J. Y. Chen, "Three-dimensional effects in CMOS latchup," in *IEDM Dig.*, p. 248, 1986.

EXERCISES

1. According to the lumped circuit model, explain why a beta product of greater than one is the necessary but not sufficient condition for latchup. Under what special case does it become the necessary and sufficient condition?

2. For a *p*-well CMOS, draw a lumped equivalent circuit model and derive conditions for latchup to occur.

3. In an *n*-well CMOS, the well depth is 1 μm and the concentration is 1×10^{16} cm^{-3}; the p^+ diffusion is 0.3 μm deep and it is doped at 1×10^{19} cm^{-3}. Estimate the vertical *pnp* transistor common emitter current gain using one-dimensional analysis.

4. Based on Fig. 8.3 for the conventional well case, calculate αnpn as a function of the n^+-to-*n*-well spacing, which is 40% of the n^+-to-p^+ spacing. Assuming well resistance (R_W) is 1.4 KΩ, calculate n^+-overvoltage induced trigger current as a function of n^+-to-p^+ spacing in the range of 5 to 12 μm.

5. Explain why latchup cannot be sustained if the holding voltage is greater than the supply voltage and what is the most effective way to increase the holding voltage.

6. If the n^+-to-*n*-well is 3 μm, the *n*-well depth is 1.3 μm, and the p^+ diffusion is 0.3 μm, estimate the minimum pulse width needed to trigger latchup.

7. The use of retrograde well improves latchup immunity in the form of increasing trigger current. Explain why such an improvement is most obvious when triggering is done at the diffusion outside the well. If a conventional *n*-well is changed to a retrograde *n*-well with the well resistance decreased by a factor of ten and the associated *pnp* current gain decreased by a factor of 5, calculate how much increase can be expected in the trigger current if n^+ overvoltage triggering is applied and how much increase if p^+ overvoltage triggering is used?

8. Explain why using a lightly-doped epitaxial layer on a heavily-doped substrate is the most effective way to increase latchup resistance. Why is a thinner epi even more helpful and what limits the continuous scaling of epi thickness?

9

CMOS DESIGN RULES

This last chapter of the book will discuss design rules associated with current and future CMOS technologies. A set of design rules is an important output produced by a technologist for circuit designers. Because design rules are derived from device characteristics and process capabilities, this chapter will first summarize the device results discussed in previous chapters, then it will discuss the effects of particular process(es) on individual design rules. Next introduced will be lambda-based design rules, which are basically technology independent. The limitation of the lambda rules will be discussed and followed by the method of modification. Examples will be given for 1.2 μm CMOS design rules. Finally, CMOS devices and technologies are summarized and future perspectives are given.

9.1 DESIGN RULES DERIVATION

Design rules are a set of numbers that describe the minimum dimension of layout geometries for each mask and the relationship, in terms of overlap and separation, among these geometries. In a broad view, design rules often include the process and electrical parameters that are important in determining circuit performance. Within this context, design rules are derived from electrical performance and processing capabilities. The electrical performance at a given dimension depends on the device characteristics of a particular technology, CMOS in this case. On the other hand, processing capabilities primarily depend on wafer fabrication capability rather than technology type (NMOS, CMOS, or bipolar).

9.1.1 Electrical Performance

Intra- and inter-device issues for CMOS technology have been described in previous chapters. This chapter will take the opportunity to summarize the results relevant to design rules. For example, the design rule of n^+-to-n^+ or p^+-to-p^+ separation not only depends on lithographic capability, but also on device physics, as punchthrough occurs between the two diffusions when they are located too close.

Transistor design and the associated parasitics are discussed in Chapter 6. For 1 μm nMOSFETs, the LDD structure should be used to alleviate the hot-electron effect. As discussed in Chapter 6, a 10-year transistor lifetime can be expected using a carefully designed LDD structure. Due to the buried-channel behavior in pMOSFETs, attention should be paid to the subthreshold swing and threshold lowering. Typical electrical parameters for 2 μm and 1.2 μm CMOS technologies are shown in Table 9.1.

All parameters presented in the table describe individual components: transistors, resistors, and capacitors. Electrical characteristics among active devices are equally important, especially in setting up inter-device design rules such as separations between two transistors. For VLSI circuits, it is desirable to have transistors packed closely together, however minimum spacings must be used so that adequate isolation among the transistors is accomplished. The isolation problem is complicated in CMOS because of two different transistor types. Various parasitic FETs exist among active transistors, and these FETs must not be turned on due to field inversion or punchthrough. As described in the last section of Chapter 7, for a standard LOCOS-isolated 1.2 μm CMOS technology, isolation among similar-type transistors can be achieved with 1 μm nominal separation. The separation is 1 μm on the mask, but ends up being about 2 μm due to approximately 0.5 μm LOCOS encroachment (bird's beak) on each side. Isolation among opposite-type transistors is more involved, however modelling and measurement results show that it is possible to achieve isolation between an n- and a p-channel transistor with approximately 2.4 μm separation on-mask. The use of retrograde wells can further reduce this dimension to below 2 μm. At such small isolation dimensions, parasitic bipolar devices, and hence latchup, become major problems.

As discussed in Chapter 8, latchup susceptibility is a strong function of n^+-to-p^+ separation and actual layout. General guidelines for designing latchup resistant circuits include large n^+-to-p^+ separation, multiple well and

TABLE 9.1 Typical Electrical Parameters for 2 μm and 1.2 μm CMOS technologies

Electric Parameters	2 μm Technology	1.2 μm Technology
Transistor Parameters:		
Gate oxide thickness	300 ± 15 Å	210 ± 10 Å
Width/length	20 μm/2.0 μm	20 μm/1.2 μm
n-channel threshold (V_{tn})	0.9 ± 0.1 V	0.7 ± 0.1 V
gain factor ($K_n = \mu_n C_{ox}$)	55 ± 6 mS/V	69 ± 6 mS/V
subthreshold swing (S_t)	90 ± 10 mV/dec	90 ± 10 mV/dec
body effect coefficient (γ)	0.20 V$^{1/2}$	0.24 V$^{1/2}$
drain breakdown (V_{BD})	25 V	11 V
p-channel threshold (V_{tp})	-0.9 ± 0.1 V	-0.8 ± 0.1 V
gain factor ($K_p = \mu_p C_{ox}$)	20 ± 4 mS/V	27 ± 4 mS/V
subthreshold swing (S_t)	110 ± 10 mV/dec	100 ± 10 mV/dec
body effect coefficient (γ)	0.45 V$^{1/2}$	0.45 V$^{1/2}$
drain breakdown (V_{BD})	-25 V	-8 V
Resistance Parameters:		
N^+ diffusion sheet resistance	35 ± 5 Ω/□	32 ± 5 Ω/□
N^- diffusion sheet resistance	— (no LDD)	1000 ± 100 Ω/□
P^+ diffusion sheet resistance	80 ± 20 Ω/□	120 ± 15 Ω/□
Poly sheet resistance	30 ± 5 Ω/□	22 ± 2 Ω/□
Polycide	3.4 ± 0.5 Ω/□	3.4 ± 0.5 Ω/□
N-well sheet resistance	2000 ± 300 Ω/□	3000 ± 300 Ω/□
Metal 1	0.032 ± 0.004 Ω/□	
Metal 2	0.039 ± 0.003 Ω/□	
Metal to n^+ contact resistance	$3\text{--}5 \times 10^{-7}$ Ω-cm^2	
Metal to p^+ contact resistance	$7\text{--}10 \times 10^{-7}$ Ω-cm^2	
Metal to n^+-poly contact resistance	$2\text{--}4 \times 10^{-7}$ Ω-cm^2	
Capacitance parameters:		
Gate capacitance in 10^{-4} pF/μm^2	11.8 ± 0.6	20.2 ± 0.65
N^+ junction area capacitance* in 10^{-4} pF/μm^2	1.0 ± 0.1	1.5 ± 0.2
P^+ junction area capacitance* in 10^{-4} pF/μm^2	2.1 ± 0.3	3.7 ± 0.5
N^+ junction sidewall capacitance* in 10^{-4} pF/μm	1.7 ± 0.2	5.6 ± 0.8
P^+ junction sidewall capacitance* in 10^{-4} pF/μm	2.5 ± 0.3	6.0 ± 0.3
Poly over field capacitance in 10^{-4} pF/μm^2	0.63 ± 0.06	0.77 ± 0.07
Metal 1 over field capacitance in 10^{-4} pF/μm^2	0.31 ± 0.02	0.35 ± 0.02
Metal 1 over poly or diffusion in 10^{-4} pF/μm^2	0.46 ± 0.04	0.49 ± 0.05
Metal 2 over field capacitance in 10^{-4} pF/μm^2	0.20 ± 0.02	0.22 ± 0.02
Metal 2 over poly in 10^{-4} pF/μm^2	0.29 ± 0.02	0.30 ± 0.02
Metal 2 over metal 1 in 10^{-4} pF/μm^2	0.68 ± 0.04	0.68 ± 0.04

*At zero volt bias.

substrate contacts, guard bands in I/O circuits, and attention to narrow-width effects. However, precise design requirements in avoiding latchup are technology- and application-dependent. Often, a latchup-free design rule is not well defined. The only definite electrical parameter in characterizing latchup is the holding voltage, which should be above the supply voltage if absolute latchup immunity is sought. This stringent requirement can be met when an epitaxial layer is used on a heavily-doped substrate. Furthermore, with an epi layer, the n^+-to-p^+ design rule can be scaled with epi thickness. For a typical 1.2 μm CMOS technology commonly configured with diffused wells, a 4 μm nominal (as grown) epi thickness is about the thinnest achieved. With this epi thickness, a 4 μm n^+-to-p^+ separation is needed to achieve a holding voltage of greater than 5 V. With retrograde-well technology, nominal epi thickness and the n^+-to-p^+ separation can both be reduced to 2 μm for V_{HOLD} > 5 V. Without an epitaxial layer, the condition for latchup immunity (V_{HOLD} > 5 V) may not be obtained even with a very large (10–20 μm) n^+-to-$p+$ separation. Depending on specific circuit applications and the use of guard bands in I/O circuits, the n^+-to-p^+ design rule may be established to provide reasonable latchup protection, even though the corresponding V_{HOLD} is <5 V.

9.1.2 Process Constraints

Pattern resolution and overlying alignment rely basically on lithographic tools. In general, an MOS technology is qualified by the minimum feature size (**f**) that can be imaged on a wafer. For example, the lithographic tool for a 2 μm (or 1.2 μm) CMOS technology is capable of producing 2 μm (or 1.2 μm) lines and separations in photoresist. As described in Chapter 3, the alignment accuracy of a lithographic tool refers to how accurately one pattern can be overlaid on top of the layer previously patterned.

Overlay errors normally result in a Gaussian distribution with the 3σ point (where σ is the standard deviation) indicating the alignment error. For most lithographic tools, it is approximately half of the resolution limit, e.g., $\pm 1\mu$m for 2 μm technology and ± 0.6 μm for 1.2 μm technology. The full resolution capability is achieved on a flat wafer. During the middle of the process, the wafer topography causes thickness non-uniformity in photoresist spun on the wafer, thus reducing the effective resolution. The resolution also worsens when imaging is done on a highly reflective layer, such as metal. For these two reasons (topography and reflection), the ultimate photoresist resolution varies from layer to layer. Furthermore, depending on the pattern transferring method, different layers require different resist thicknesses and, in general, a thicker resist layer produces poorer resolution.

Pattern transferring from the imaging (photoresist) layer to the underlying material is done by etching or implantation followed by diffusion. Other process steps, such as LOCOS or thin film deposition, can also impact design rules directly or indirectly.   For example, modified LOCOS oxidation may result in a reduced bird's beak, hence tighter design rules for the active area definition.   A layer of polysilicon deposited with poor uniformity may cause larger overetch during poly gate definition, an example of the indirect effect. As described in Chapter 5, a typical CMOS process consists of at least 10 masking layers: well definition, active area, poly gate, p^+, n^+, contacts, metal 1, via, metal 2, and pad.   The process steps for making each layer are different and design rules for each layer should be individually derived.

Well definition.   Wells are normally formed by diffusion, which makes the actual well larger than what is drawn on the mask and imaged in the resist.   In fact, the lateral extension of a well at one edge is approximately 75% of the well's depth.   Therefore, minimum well width (W_w) is the sum of minimum feature size and twice the lateral extension.   The separation between two wells at two different potentials is limited by electrical isolation. To prevent punchthrough in the lightly-doped substrate between the two wells, adequate spacing between them must be used.

Active area.   Active areas are the regions where active transistors are built.   In a standard MOS process, active area is defined by a layer of nitride mask followed by LOCOS oxidation.   As discussed in Chapter 7, the actual active area is smaller than what appears on the mask due to the oxidation encroachment known as bird's beak.   In a conventional LOCOS process, bird's beak is roughly 80% of field oxide thickness.   For a 0.6 μm field oxide, roughly a 0.5 μm retreat at each edge of the active area exists.

Poly gate.   Poly gates are formed by etching a layer of doped polysilicon.   Because the poly linewidth corresponds to the FET gate length, its dimension is the minimum feature size (**f**).   With an anisotropic RIE (Reactive Ion Etching) technique, essentially no dimension variation is found when the pattern is transferred from resist to the poly layer.   The poly gate must extend beyond the transistor's active area to avoid short circuit between source and drain.

p^+ and n^+.   These two masks are needed to form p- and n-channel source/drain regions.   Using a photoresist mask and the existing poly gate, boron or arsenic ions are implanted only for the p- or n-channel S/D areas, respectively.   To account for alignment error, the resist mask is normally designed to overlap the active area so that the entire p- or n-channel S/D regions are implanted.   When an LDD structure is included, n^- is formed using a phosphorus implant with or without a resist mask.   In the case of no mask, phosphorus impurities in the p-channel S/D region are compensated by the much heavier boron implant dose.

Contacts. Contacts are openings in the oxide layer down to the poly gates or Si substrate. To make them, holes are patterned in a photoresist layer and subsequently transferred to the underlying oxide using RIE etching. Several process constraints exist. First, minimum size ($\mathbf{f} \times \mathbf{f}$) contacts are normally required for saving layout area. To image a minimum square is often much more difficult than printing a line with minimum width. Secondly, sloped contacts may be more desirable to ease the step coverage problem often encountered during subsequent metal deposition. This requirement creates a heavier burden in the etching process. Finally, a finished contact must be inside an active area or a poly gate. For this reason, one end of a poly gate is always abutted to a poly pad. Moreover, when the contact mask is aligned to the active (or poly) mask, the relative mis-alignment between contacts to poly (or active area) can be much larger than the overlay alignment error.

Metal 1. Metal lines are used to connect transistors through contacts. Metal lines are patterned and etched in a similar way to the formation of poly lines. Minimum linewidths and separations are desirable for high packing density. Metal pitch, defined as the sum of linewidth and separation, is often the determining factor in VLSI layout size. Imaging high-resolution metal patterns is more difficult because of the high reflecting metal layer. In addition, metal lines must be wide enough to cover the contacts. Thus, minimum metal linewidth is often set by the minimum contact size plus the amount of overlap between the metal and the contact.

Via and metal 2. Vias and metal 2 lines are formed similarly to contacts and metal 1 lines. Vias are etched down to the first level metal instead of Si substrate. Directly contacting Si through vias in general is not allowed. Because wafer topology is uneven at this point, resolution for via and metal 2 is poorer. As a result, the corresponding design rules are often looser than those for contact and metal 1 unless planar techniques are used to provide planar surface after metal 1 formation.

Pad. ICs normally require a layer of dielectric (oxide or nitride) at the end of the process for passivation. A pad mask is used to open a large hole on bonding pads for probing and bonding. The size (75–100 μm on each side) is far larger than what can be made, hence no specific issues arise relating to minimum design rules.

9.2 LAMBDA-BASED DESIGN RULES

As previously described, pattern resolution and alignment accuracy improve together as the lithographic tool progresses. Other process techniques such as etching also improve with technology advances. For these reasons, design rules have been developed in dimensionless form in terms of the length unit lambda (λ), which is a time-varying parameter.[1]   These rules assume certain linear relationships among various layout elements and greatly ease the design changes through λ shrinkage as the technology advances.   The assumption is that all rules improve at the same rate, hence the design rules themselves need not be altered when they are expressed by λ. λ rules have been used successfully in designing NMOS ICs.   They have also been applied to CMOS,[2-4] even though the implementation is less straightforward.   The λ reduction works fairly well at the 2-to-3 μm range because the linear relationships more or less hold.   They can also be stretched to 1.2 μm design provided tight layout is not critical compared to the layout time; examples include gate arrays, standard cells, and semi-custom ICs.

Based on λ rules, the minimum imaging size (f) is usually 2λ and alignment accuracy, although dependent on the lithographic machine, is roughly the λ value.   The CMOS layering sequence and corresponding intra-layer λ design rules for 1–2 μm technologies are shown in Table 9.2.   The mask alignment sequence, which indicates the previous layer for each layer to align to, is also included in the table.   Most, but not all of the layers, are aligned to the layers that are just one layer preceding them.   For example, Layer 3 (Poly Gate) is aligned to Layer 2 (Active Area), and Layer 2 is aligned to Layer 1 (Well), but Layer 6 (Contact) is aligned to Layer 3 (Poly Gate), not Layer 5.

TABLE 9.2   CMOS Layering Sequence and the Corresponding Intralayer Lambda Design Rules for the 1-2 Micron Technologies

Layer No./Name	Minimum Feature, λ	Minimum Separation, λ	Alignment Sequence
1. Well (WE)	6	9	—
2. Active Area (AA)	2	3	Well
3. Poly Gate (POL)	2	2	Active area
4. n+ (N+)	3.5	—	Active area
5. p+ (P+)	3.5	—	Active area
6. Contact (CT)	2	2	Poly gate
7. Metal 1 (MET1)	3	3	Contact
8. Via (VIA)	2	3	Metal 1
9. Metal 2 (MET2)	3	4	Via

The λ values equal 1 μm for a 2 μm CMOS technology and 0.5 μm for a 1 μm technology. Notice that the minimum features, such as poly gates and contacts (or vias), have 2λ size. These features are critical dimensions and have to be held precisely. If the poly gates are oversized, the current driving ability degrades; if they are undersized, then short-channel effects in the form of threshold lowering and increased punchthrough current are more pronounced. Contacts are made of holes in an insulator layer. If the finished contact holes are smaller than the designed size, the corresponding contact resistance increases, which tends to decrease current and can even cause open circuit in extreme cases. An oversized contact hole can cause shorts and may not be completely covered by metal lines. With hundreds of thousands of contacts in a VLSI circuit, contact failure is a major yield consideration. IC designs often use single sized (2λ $\times$ 2λ square) multiple contacts instead of one large sized contact for a larger S/D diffusion area. Lithographic and etching processes can be optimized for the best control of the only one sized contact. The minimum well width and separation are much larger for reasons described in the last section. Lines or separations for the first-level metal (metal 1) are 3λ, resulting in a metal 1 pitch of 6λ. Metal 2 lines are also 3λ with 4λ separations between the two lines. Because n^+ and p^+ must overlap the active area, their minimum feature size is 3.5λ.

Inter-layer design rules, such as overlaps and separations, depend on lithographic misalignment and pattern transfer variation, such as the amount of overetch. When each layer is aligned to a previous layer according to the mask alignment sequence, the alignment error between the two layers is fixed by the lithographic machine. But, the alignment error between any two layers is not the same. For example, the Contact-to-Poly misalignment Δ_{CT-PL} is about the same as the Poly-to-Active misalignment Δ_{PL-AA}, but is different from the Contact-to-Active misalignment Δ_{CT-AA} due to the accumulation of alignment errors. The definition of the misalignment (Δ) at the 3σ point means that 99.73% of the alignment errors fall in $\pm\Delta$. Because the Contact layer is aligned to the Poly-Gate layer, which is aligned to the Active-Area layer, the Contact-to-Active misalignment (Δ_{CT-AA}) is statistically larger than Δ_{PL-AA} or Δ_{CT-PL}. More specifically, because both Poly-to-Active and Contact-to-Poly alignment errors follow Gaussian distribution, the misalignment between Contact to Active is also a Gaussian distribution with its 3σ point being $\sqrt{2}$ times the 3σ of the Poly-to-Active or Contact-to-Poly alignment errors. If $\Delta_{PL-AA} = \Delta_{CT-PL} = 0.5$ μm, $\Delta_{CT-AA} = \sqrt{2} \times 0.5$ μm $= 0.7$ μm, meaning that 99.73% of the alignment errors between Contact to Active are within ±0.7 μm. The accumulation of these alignment errors should be taken into account when defining overlaps and separations for the inter-layer design rules. The linewidth variation during pattern transferring varies depending on the process of each layer and must be obtained individually.

Table 9.3 shows typical CMOS λ design rules offered by MOSIS (MOS Implementation System). The numbers shown in the table are in units of λ.

MOSIS CMOS SCALABLE RULES

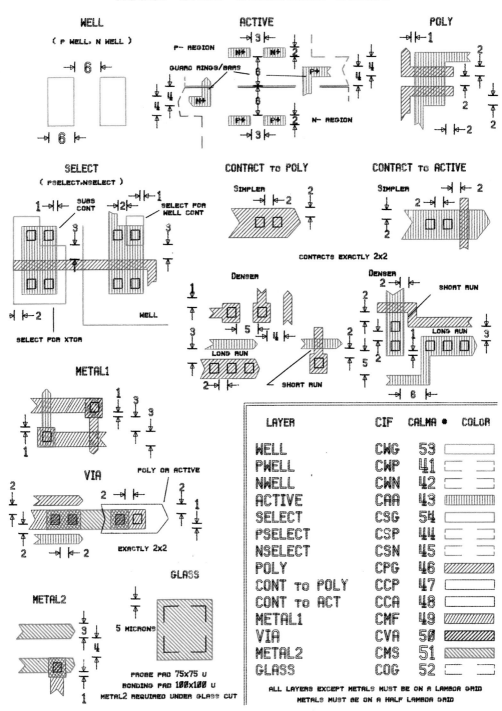

Features are drawn on a λ grid except metals, which stay on a $\lambda/2$ grid. λ is 1.5 μm for 3 μm technology, 1.0 μm for 2 μm technology, and 0.7 μm for 1.2 μm technology. Process biases (compensations) are done at the mask making step.

For latchup protection, the n^+-to-p^+ separation is 12λ, with the n^+-or-p^+ source/drain to well edge being 6λ. For 1.2 μm technology, this rule gives a 8.4 μm n^+-to-p^+ separation, larger than the minimum 4 μm separation previously discussed. Thus, thicker epi layers or even bulk substrates can be used. The minimum separation between similar active areas (n^+-to-n^+ or p^+-to-p^+) is 3λ as drawn, but is reduced on the mask to compensate the bird's beak effect during fabrication. The extension of the Poly Gate beyond the Active Area is drawn 2λ to ensure a finite overlap in finished Si wafers at the worst case of misalignment and poly overetch. NSELECT is used for n^+ source/drain regions and n-well contact. Similar usage applies to PSELECT. Again, 2λ overlap is established to ensure that the active areas are covered. Contacts are divided into contacts to poly and contacts to active so that the poly layer or active layer or just the overlap around the contacts can be adjusted independently. They are also divided into simpler and denser contacts. Minimum overlaps of active or poly to simpler contacts are 2λ whereas minimum overlaps of active or poly to denser contacts are 1λ. Contacts or vias are at least 2λ away from the unrelated active or poly layers so that finite separations are preserved. Overlaps of metal 1-to-contact and metal 2-to-via are 1λ in both cases.

9.3 LIMITATION OF LAMBDA RULES AND THEIR MODIFICATIONS

Lambda design rules are convenient for designers dealing with different technologies from time to time. However, the accuracy in positioning and transferring patterns does not always scale uniformly for all the layers when technology evolves. Subtle technology differences may also exist among various chip makers. Modifications in λ rules may be necessary to accommodate these differences. One way to handle this problem so that it is basically transparent to the designer is to apply process-dependent biases during the mask making operation. By doing so, the final dimensions in Si are in principle equal to what were drawn by the designer. Features or overlaps can be bloated or shrunk by different amounts for different layers. The bloat or shrink can also be independent of technology in terms of minimum feature size, e.g, 1.2 or 2.0 μm technology. By doing so, the spirit of λ rules can be extended to 1.2 μm technologies and the same design can be fabricated by different chip makers.[6] Table 9.4 shows an example of modified λ design rules established for a 1.2 μm CMOS technology.

9.4 SUBMICRON DESIGN RULES AND THEIR IMPACTS

As IC technology evolves into the submicron regime, λ-based design rules may not be applicable, because the relationships among the various rules change significantly. For example, variation in linewidth control for each layer improves differently due to different process steps, e.g., LOCOS in the active layer versus etching in the poly layer. Therefore, when applying λ-scaling to layout at submicron dimensions, one layer may reach its process limit before the others do resulting in a relaxed layout with low packing density. In particular, for memory designs in which every micron counts, λ rules do not apply. Furthermore, design rules established by electrical performance may not scale with λ because of enhanced short- and narrow-channel effects. Minor modifications of λ rules using the above-mentioned bloat and shrink may not be sufficient to provide an optimal layout. It is perhaps more adequate to work on design rules directly expressed by microns.

As far as submicron CMOS design rules are concerned, optical lithography has pushed the resolution limit to 0.8 μm or below, allowing a 4-Mbit DRAM to be made on a single chip. Using 0.8 μm design rules, 4 Mb DRAM chips have been made with chip sizes around 100 mm^2 so that conventional packages can be used.[7-9] At 5 V supply, these 4 Mb DRAMs have a typical access time of 65–95 ns. 0.8 μm CMOS design rules have also been applied in 1 Mb SRAMs with chip sizes between 80–100 mm^2.[10-12] Address access times for these 1 Mb SRAMs range from 15–25 ns with a 5 V supply. An LDD structure is used at least for the n-channel transistors. Table 9.5 shows the 0.8 μm design rules and corresponding performances for the Hitachi double-poly double-Al 1 Mb CMOS SRAM.[12] The poly-Si and Al pitches are approximately 2 μm with the contact holes at 0.8 μm. The corresponding layout sizes and performances are also shown.

Design rules are being extended to a 1/2 μm dimension without being limited by performance saturation; again, lambda rules do not apply. As 0.5 μm design rules are approached, the electrical fields are so high that hot-carrier induced degradation occurs even in p-channel transistors. Consequently, LDD may have to be applied to both types of devices unless the power supply can be reduced. With a 3.3 V supply voltage, a 16 Mb DRAM has been demonstrated using 0.7 μm CMOS design rules, including electron-beam direct writing.[13] The effective channel lengths are 0.5 and 0.7 μm for nMOS and pMOS transistors, respectively, and the gate oxide is 120 Å for both types of devices. A cell size of 5 μm^2 is achieved using self-aligned silicided interconnection. The total chip size is about 150 mm^2. Similar design rules and technology have also been applied in fabricating 0.5 μm 32b CMOS macros.[14] The effective channel lengths are 0.4 and 0.5 μm for n- and p-channel FETs, respectively. An **ADD** speed of 8 ns in the macro-chip 32b ALU has been measured.

TABLE 9.4   Modified Lambda Design Rules for 1.2 Micron CMOS

	As Drawn (unit = λ)	As Drawn (unit = μm)	On Mask (unit = μm)
Active Area (AA) (0.4 μm bloat per side)			
AA	2	1.4	2.2
AA–AA separation	3	2.1	1.3
AA (p^+-n^+)	10	7.0	6.2
AA beyond POL	3	2.1	2.5
AA-POL (opposite type)	3	2.1	1.7
Poly (POL) (no bloat or shrink)			
POL	2	1.4	1.4
POL-POL	2	1.4	1.4
POL-AA (unrelated)	1	0.7	0.3
POL extend beyond AA	2	1.4	1.0
POL (minimum width FET)	3	2.1	2.9
Contact (CT) (0.1 μm shrink per side)			
CT	2	1.4 × 1.4	1.2 × 1.2
CT–CT separation	2	1.4	1.6
AA overlap CT	1	0.7	1.2
POL overlap CT	1	0.7	0.8
CT (in AA)-POL	2	1.4	1.5
CT (in POL)-AA	2	1.4	1.1
Metal 1 (MET1) (0.2 μm bloat per side)			
MET1	3	2.1	2.5
MET1–MET1 separation	3	2.1	1.7
MET1 overlap CT	1	0.7	1.0
MET1 overlap VIA	1	0.7	1.0
MET1 Pitch	6	4.2	4.2
VIA (0.1 μm shrink per side)			
VIA	2	1.4 × 1.4	1.2 × 1.2
VIA-VIA	3	2.1	2.3
VIA-POL	2	1.4	1.5
VIA-AA	2	1.4	1.1
Metal MET2 (0.2 μm bloat per side)			
MET2	3	2.1	2.5
MET2-MET2	4	2.8	2.4
MET2 overlap VIA	1	0.7	1.0
MET2 Pitch	7	4.9	4.9
WELL (no bloat or shrink)			
WELL	6	4.2	4.2
WELL-WELL	6	4.2	4.2
WELL overlap p^+	6	4.2	4.2
WELL-n^+	6	4.2	4.2
p^+-n^+	12	8.4	8.4

TABLE 9.5 Eight-Tenths Micron CMOS Design
Rules for 1 Mbit Hitachi SRAM (Ref. 12, © 1988
IEEE).

Technology	0.8 μm CMOS
	Double poly-Si, Double Al
Gate length	0.8 μm (nMOS and pMOS)
Poly-Si (width/space)	0.8 μm / 0.8 μm
Contact hole	0.8 μm
1st Al (width/space)	1.4 μm / 0.8 μm
Via hole	0.8 μm
2nd Al (width/space)	1.4 μm / 0.8 μm
Organization	256K word × 4 bit
Chip size	6.15 × 15.21 mm²
Cell size	5.2 × 8.5 μm²
Address access time	15 ns
Operating current	50 mA at 20 MHz
Standby current	2 μA
Package	28 pin, 400 mil SOJ
Redundancy	1 row and 2 columns

Scaling CMOS design rules to $\leq 1/4$ μm for VLSI circuits faces more challenge due to the diminishing returns on performance improvement, increasing reliability concerns, practical limitations in manufacturing, and eventual fundamental limits on device physics. Three-dimensional or hybrid integration (such as BiCMOS/CMOS or GaAs/Si) and advanced packaging offer other opportunities for VLSI circuits and systems.

9.5 SUMMARY AND FUTURE PERSPECTIVES

With its inherent low power characteristics and advantages in circuit design, CMOS is undoubtedly the dominant VLSI technology. In the next decade, most digital designs, including microprocessors and memories, will be switched from NMOS to CMOS. More and more analog circuits will be designed in CMOS to replace bipolar circuits. In VLSI, only CMOS circuits can run at high speed without excessive heat generation. For high density and speed VLSI applications and/or mixed analog/digital circuits, BiMOS (bipolar/CMOS) will gain increasing attention. The selection of a particular CMOS technology will depend largely on circuit applications and, to a lesser degree, technology evolution. For one-to-two micron designs, n-well technology seems more appropriate for many VLSI circuits, because n-channel devices perform better, and generally more of them are used in a circuit. Moreover, it is natural to extend n-well CMOS to BiCMOS. As the transistor gate length shrinks below one micron, twin-tub and retrograde well technology will become more attractive.

The key device issues in CMOS VLSI are hot-electron effects in n-channel transistors, buried-channel behavior in p-channel transistors, isolation between n and p transistors, and latchup. Electrical parameters in setting up the design rules are strongly dependent on these key issues. LDD has been used to reduce the hot-electron effect. Care must be exercised in LDD design and process to achieve desired reliability at a minimal decrease of circuit speed. The hot-electron-induced degradation and other reliability concerns, such as thin oxide quality and electromigration, may place fundamental limits in scaling CMOS devices to the submicron regime.[15] Punchthrough current is critical in designing pMOS due to its buried-channel nature. Both low threshold and low off-state leakage must be simultaneously obtained. As the effective channel length is reduced to half-micron or below, pMOS also suffers from hot-carrier degradation unless the 5 V supply can be reduced. In CMOS isolation, both parasitic FETs and bipolars must be considered to prevent field inversion, punchthrough and latchup. The isolation problem is complicated in CMOS because of two different transistor types and many p-n junctions. It becomes difficult when devices are packed closely together, as required in VLSI circuits. With retrograde wells and epitaxial substrates, high density ICs with 2 micron n-to-p-channel separations can be realized with good isolation, including latchup immunity. For high-density CMOS circuits, such as mega-bit SRAMs, trench isolation or selective-epi growth will be ideal.

For half-micron CMOS technology, parasitic resistance and capacitance must be reduced with channel length reduction so that circuits can benefit from scaling. For example, new technology, such as self-aligned silicidation, is being developed to reduce the parasitic source/drain resistance by covering the entire source/drain (S/D) area as well as the gate with metal silicide. This technology is particularly important for CMOS because p^+ resistance is high for shallow p-channel S/D. In VLSI, interconnect technology is another important aspect, especially for CMOS, because multiple input gates require more connections and because of the lack of buried contacts commonly used in NMOS. Two-level interconnect is necessary and three-level wiring may not be far in the future.

New technologies, such as three-dimensional ICs, may be more effective in packing more transistors on a chip or achieving a higher level of integration. It has already been seen, for example, that a 64K CMOS SRAM can be fabricated by making p-channel load devices in a hydrogen-passivated poly-Si layer on top of n-channel transistors made in the substrate.[16,17] A double-layer IC with high performance devices in both layers (Si substrate and a laser-recrystallized SOI layer) has been demonstrated via a 256-bit SRAM[18] as the first step to 3-D ICs. CMOS, again due to its low power dissipation, will be the dominant technology in 3-D VLSI circuits.

REFERENCES

1. Mead and Conway, *Introduction to VLSI System*, Addison-Wesley, 1980.

2. T. W. Griswoid, "Portable design rules for bulk CMOS," *VLSI Design*, p. 62, Sept/Oct., 1982.

3. M. R. Gulett, "Scaling down CMOS design rules," *VLSI Design*, p. 24, Sept., 1983.

4. J. Y. Chen, "Scaling CMOS to submicron design rules for VLSI," *VLSI Design*, p. 78, 1984.

5. G. Lewicki, "MOSIS Design Rules," Information Science Institute.

6. J. D. Trotter, "Scalable CMOS design rules," Mississippi State University, private communication.

7. M. Takada, T. Takeshima, M. Sakamoto, T. Shimizu, H. Abiko, T. Katoh, M. Kikuchi, S. Takahashi, Y. Sato and Y. Inoue, "A 4Mb DRAM with half internal-voltage bitline precharge," *ISSCC*, p. 270, 1986.

8. K. Mashiko, M. Nagatomo, K. Arimoto, Y. Matsuda, K. Furutani, T. Matsukawa, T. Yoshihara, and T. Nakano, "A 90 ns 4Mb DRAM in a 300 mil DIP," in *ISSCC Dig.*, p. 12, 1987.

9. K. Shimohigashi, K. Kimura, Y. Sakai, H. Tanaka, and K. Yagi, "A 65 ns DRAM with a twisted driveline sense amplifier," in *ISSCC Dig.*, p. 18, 1987.

10. T. Ohtani, K. Hashimoto, M Matsui, J-I Tsujimoto, J. Iwai, M. Saitoh, H. Shibata, H. Sasaki, M. Isobe, J-I. Matsunaga, and T. Iizuka, "A 25 ns 1Mb CMOS SRAM," in *ISSCC Dig.*, p. 264, 1987.

11. T. Wada, T. Hirose, H. Shinohara, Y. Kawai, K. Yuzuriha, Y. Kohno, Y. Sakai, S. Meguro, M. Tsunematsu, and T. Masuhara, "A 34 ns 1Mb CMOS SRAM using triple poly," in the *ISSCC Dig.*, p. 262, 1987.

12. K. Sasaki, S. Hanamura, K. Ueda, T. Onono, O. Minato, K. Nishimura, Y. Sakai, S. Meguro, M. Tsunematsu, and T. Masuhara, "A 15 ns 1Mb CMOS SRAM," in the *ISSCC Dig.*, p. 174, 1988.

13. T. Mano, T. Matsumura, J. Yamada, J Inoue, S. Nakajima, K. Minegishi, K. Miura, T. Matsuda, C. Hashimoto, and H. Namatsu, "Circuit technologies for 16Mb DRAMs," in the *ISSCC Dig.*, p. 22, 1987.

14. C.L. Chen, L.K. Wang, A. Edenfeld, and P. Nixon, "Two CMOS 0.5 μm 32b digital Macros," in the *ISSCC Dig.*, p. 62, 1987.

15. M. H. Woods and B. L. Euzent, "Reliability in MOS integrated circuits," in *IEDM Tech. Dig.*, p. 50, 1984.

16. P. K. Chatterjee, "Device design issues for deep submicron VLSI," *Proc. of International Sym. on VLSI Technology, Systems and Applications*, Taipei, Taiwan, p. 221, 1985.

17. H. Shichijo, S. D. Malhi, W. F. Richardson, G. P. Pollack, A. H. Shah, L. R. Hite, S. K. Banerjee, M. Elahy, R. Sundaresan, R. H. Womack, H. W. Lam and P. K. Chatterjee, "Polysilicon transistors in VLSI MOS memories," in *IEDM Tech. Dig.*, p. 228, 1984.

18. T. Nishimura, Y. Inoue, K. Sugahara, M. Nakaya, Y. Horiba and Y. Akasaka,

"A three dimensional static RAM," in the *Dig. of the Sym. on VLSI Technology*, p. 30, 1985.

EXERCISES

1. Why is parasitic capacitance per unit area smaller for the metal 1 on field oxide, but much larger for the polysilicon lines on field oxide? Based on Table 9.1, calculate field oxide thickness under the polysilicon and the field oxide thickness under the metal 1.

2. It is known that electric (or effective) channel length (L_{eff}) is different from the drawn channel length (L_{drawn}). Describe a method that one can extract L_{eff} from a series of test devices with different L_{drawn}. Explain what are the major processes responsible for ΔL, where $\Delta L = L_{drawn} - L_{eff}$.

3. If LDD structure is needed for both *n*- and *p*-channel FETs in a submicron CMOS process, list the CMOS masking sequence and associated major process steps. If doping concentration of the p^- LDD is an order of magnitude larger than that of the n^- LDD, can some of these mask levels be skipped?

4. In first order approximation, the overlay alignment error is lithographic machine dependent but is independent of a particular pair of layers as long as they are the two layers purposely aligned to each other. This is however not necessarily true because the actual alignment does depend on the material and the topography of the alignment marks on a particular layer. If the alignment errors at 3σ point for the poly-to-active is 0.7 μm and the alignment errors (again at 3σ) for the contact-to-poly is 0.4 μm, what is the misalignment between contact to active at 3σ point?

5. Based on the modified 1.2 μm lambda design rules in Table 9.4, lay out a 6-transistor CMOS SRAM cell with minimum Si area. Convert this layout onto a set of CMOS masks using bloat and shrink provided by Table 9.4.

6. Based on the MOSIS scalable design rules, lay out the CMOS buffer circuit described in Chapter 4 and estimate the layout area if 2 μm CMOS technology is used. For the driver of the buffer, include guard bands in the layout to avoid latchup.

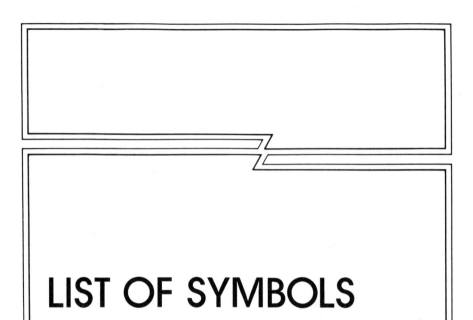

LIST OF SYMBOLS

Symbol	Description
α	bipolar common base current gain
α_n	electron impact ionization rate
α_p	hole impact ionization rate
A_E	emitter area
β	bipolar common emitter current gain
β_{npn}	common emitter current gain for an npn bipolar
β_{pnp}	common emitter current gain for a pnp bipolar
C_{ox}	oxide capacitance per unit area
C_d	depletion capacitance per unit area
C_g	gate capacitance per unit area
C_m	Miller capacitance
C_s	stray capacitance
C_L	load capacitance
C_{gd}	gate-to-drain capacitance
C_{gs}	gate-to-source capacitance
C_{gb}	gate-to-bulk capacitance
C_{ov}	overlapping capacitance
C_{fr}	fringing field capacitance
C_j	junction capacitance
D	diffusivity
D_i	intrinsic diffusivity
D_n	electron diffusion constant

Symbol	Description
d	depletion width
d_{max}	maximum depletion width
d_{inv}	inversion layer thickness
ξ_i	threshold energy for impact ionization
ξ_b	threshold energy for overcoming Si/SiO$_2$ energy barrier
E	electric field
E_m	maximum electric field
E_{eff}	effective electric field
E_{crit}	critical electric field
E_a	activation energy
E_c	conduction band edge
E_v	valence band edge
E_i	intrinsic energy level
E_F	Fermi level
E_{Fn}	quasi Fermi level for electrons
E_{Fp}	quasi Fermi level for holes
ε_{ox}	oxide dielectric constant
ε_{Si}	silicon dielectric constant
Φ_B	bulk potential
Φ_s	surface potential
Φ_{ms}	work function difference
Φ_m	work function of a metal
Φ_{se}	work function of a semiconductor
f_t	bipolar cutoff frequency
G	generation rate due to impact ionization
g_d	MOSFET channel conductance
g_m	MOSFET transconductance
I_E	emitter current
I_{NE}	electron current injected from emitter
I_{PB}	hole current injected from base
I_B	base current
I_C	collector current
I_{NC}	electron current at base-collector junction
I_{sub}	substrate current
I_d (I_D)	drain current
I_g (I_G)	gate current
I_S	source current
I_{trig}	latchup trigger current
I_{hold}	latchup holding current
J_n	electron current density
J_p	hole current density
k	Boltzman constant
λ	dimensionless unit for lamda design rules
λ_e	hot electron mean free path
l	characteristic channel length
L	MOSFET channel length
L_{eff}	MOSFET effective channel length
L_{nB}	electron diffusion length in an *npn* base
N_A	acceptor impurity density
N_D	donor impurity density

Symbol	Description
n_i	intrinsic free carrier concentration
n	free electron concentration
n_B'	excess electron density in an npn base
n_{BO}	electron density in an *npn* base at equilibrium
p	free hole concentration
p_{EO}	hole density in an *npn* emitter at equilibrium
p_E'	excess hole density in an *npn* emitter
ρ	charge density
ρ_c	contact resistivity
Q_m	total charge at MOS gate
Q_n	free charge at MOS inversion layer
Q_B	total depletion charge
Q_{fc}	fixed charge density
R	recombination and generation rate
R_d	drain resistance
R_s	source resistance
R_S	substrate resistance for latchup characterization
R_W	well resistance for latchup characterization
R_{tr}	transistor resistance
R_{SE}	emitter sheet resistance
R_{SB}	base sheet resistance
R_C	contact resistance
R_p	projected range in implant profile
γ	body effect coefficient
γ_e	emitter efficiency
σ_p	standard deviation (straggle) in an implant profile
S	MOSFET scaling factor
S_t	subthreshold swing factor
T_e	electron temperature
t_{ox}	gate oxide thickness
t_{fox}	field oxide thickness
t_r	rise time
t_f	fall time
t_{in}	inverter delay time
τ_t	bipolar base transit time
μ	mobility
μ_n	electron mobility
μ_p	hole mobility
v	carrier velocity
V_{BE}	base emitter bias voltage
V_{ben}	base-emitter voltage for an *npn* bipolar
V_{bep}	base-emitter voltage for a *pnp* bipolar
V_{bi}	p-n junction built-in voltage
V_s (V_S)	source voltage
V_d (V_D)	drain voltage
V_{dd} (V_{DD})	power supply voltage
V_{dsat}	saturation drain voltage
V_g (V_G)	gate voltage
V_{sub}	substrate voltage
V_t	MOS threshold voltage

Symbol	Description
V_{tn} (V_{Tn})	n-channel threshold voltage
V_{tp} (V_{Tp})	p-channel threshold voltage
V_{TE}	threshold voltage for an enhancement-mode FET
V_{TD}	threshold voltage for a depletion-mode FET
V_{TF}	threshold voltage for a field oxide FET
V_{FB}	flat-band voltage
V_{pt} (V_{PT})	punchthrough voltage
W	MOSFET channel width
W_{eff}	MOSFET effective channel width
W_B	bipolar base width
W_E	bipolar emitter width
x_j	junction depth
χ	electron affinity

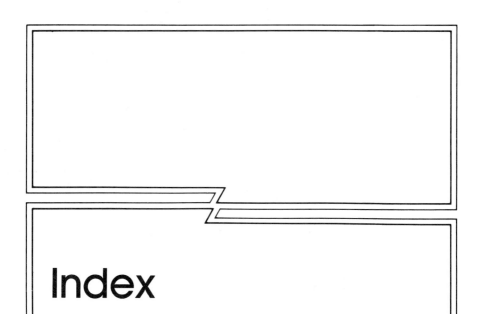

Index